Troubleshooting and Servicing Air Conditioning Equipment

Troubleshooting and Servicing Air Conditioning Equipment

S. Don Swenson

North Seattle Community College

Breton Publishers
A DIVISION OF WADSWORTH, INC.
Boston, Massachusetts

Breton Publishers
A Division of Wadsworth, Inc.

Library of Congress Cataloging in Publication Data

Swenson, S. Don, 1928–
    Troubleshooting and servicing air conditioning
equipment.

    Includes index.
    1. Air conditioning—Equipment and supplies—Mainte-
nance and repair.   I. Title.
TH7687.7.S94   1985   697.9'3'0288   84–29283

ISBN 0-534-04932-X

Printed in the United States of America
1 2 3 4 5 6 7 8 9–89 88 87 86 85

Contents

v

Chapter 3

Hazards and Safety   30

Chapter 4

Troubleshooting and Service   42

Chapter 5

Refrigeration System   48

Chapter
12

Servicing Electric Motors and Controls   161

Chapter
13

Service Procedures Common to All Refrigeration Systems   177

Chapter
14

Compressor and High-Pressure Side of the System   194

Chapter
15

Servicing the High-Pressure Side of the System   219

Chapter
16

Expansion Device and Low-Pressure Side of the System   238

Chapter
17

Servicing the Low-Pressure Side of the System   251

Chapter 21

Heat Pump Service Charts   321

Chapter 22

Checking Specific Parts of Heat Pump Systems   327

Chapter 23

Maintenance   342

Appendix Tables   352

Preface

During many years of teaching air conditioning and heating classes, I have perceived a definite need for a text that would deal with the areas of troubleshooting and servicing in greater depth than existing books. I believed that a text was needed which would review the basic theory, tools, and safety procedures and then present a systematic approach to troubleshooting. This system would then be applied to servicing common problems in each of the major components of air conditioning, heat pump, and refrigeration units. This book is an effort to provide this necessary learning material.

Assuming only that the reader has the ability to understand basic HVAC principles, this book is intended for professional and student service technicians. It can supplement any HVAC basic text or it can be used as the core text for a class focusing on troubleshooting/servicing. The material is organized to progress from general principles of safety, tools, chart interpretation, instrumentation, and troubleshooting to descriptions of specific systems and components. The last portion of the text deals with practical servicing procedures.

Some of the special features of this text include:

- **Systematic Coverage:** This text presents a general systematic method of troubleshooting air conditioning equipment without limiting its scope to specific manufacturers.
- **Focused Approach:** This text, clearly and concisely, covers all the essentials and nothing more.
- **Indispensable Charts:** Extensive troubleshooting and servicing charts provide concise summaries of the book's information as well as reference for on-the-job situations.

- **Concept Review:** Chapters 2 and 3 review basic information students must know before they begin troubleshooting and service procedures.
- **Tool and Safety Information:** Early chapters provide excellent coverage of these crucial topics.
- **Review Problems:** A large number of problems and questions are included at the end of every chapter.

The author wishes to acknowledge the editorial and production staff at Breton Publishers, without whose inspiration and guidance this project would have been much more difficult to accomplish. As the manuscript evolved, several reviewers provided technical assistance and valuable insights into the structure of the work. The author wishes to acknowledge Albert J. Buzak of Northeast Institute, Gilbert Luhmann of Long Beach City College, and Bob Arnold of Texas State Technical Institute for their assistance.

Troubleshooting and Servicing: Basic Concepts

1.1

Introduction

Air conditioning and heat pump systems are commonly used to provide a comfortable temperature in homes, stores, and places of entertainment throughout the world. There are also other applications for these systems, such as commercial cooling processes for production work and the cooling and heating of spaces where precise temperature conditions must be maintained. An example of a commercial cooling process is the cooling of computer rooms. An example of cooling and heating for maintenance of precise conditions is the cooling and heating of a "clean room" in a chemical laboratory.

If the air conditioning system in an air-conditioned building stops working, the air in the building may become uncomfortably warm, but the result of the problem is usually only inconvenience. However, if a system stops working in a commercial or a critical control system, important processes may be interrupted or even stopped. Air conditioning equipment may fail at times. All air conditioning and heat pump systems are made up of mechanical parts. If a part fails, the system will usually shut down until repairs are made.

Because of the widespread use of air conditioning and heat pump systems—and the need for their proper operation—trained technicians are necessary to find problems in the systems when they occur and to correct the problems as quickly as possible. Checking a system to find out why it isn't functioning correctly is the job

Troubleshooter
Service technician

of a **troubleshooter.** Correcting the problem is the job of a **service technician.**

In this chapter, air conditioning and heat pump systems are briefly described, the processes of troubleshooting and service are discussed, and the procedures used in troubleshooting and servicing a system are previewed.

1.2

Description of Air Conditioning and Heat Pump Systems

Air conditioning system
Heat pump system

An **air conditioning system** is used for cooling air, while a **heat pump system** is used to both cool and heat air. In this text, the general term *air conditioner* is used to describe both systems when the refrigeration part of the system or a cooling function is discussed. The term *heat pump* is used when a unit that provides both cooling and heating is described.

There are three general types of air conditioning systems: comfort cooling systems, production cooling systems, and critical-environment control systems. Each type is described in the following paragraphs.

Comfort cooling systems

Comfort cooling systems include systems that are used in residences and places of entertainment or work, where the main concern is temperature control to maintain a comfortable atmosphere. In warmer climates, such air temperature control is necessary for comfort. But even in cooler climates, the internal heat loads in a building can easily create a need for air conditioning.

Production cooling systems

A **production cooling system** is one in which temperature control is important to the processes carried on in a room or when the air temperature is important to the material being processed. Many material handling processes and some manufacturing processes require such control. For example, air conditioning equipment is used for process control in computer rooms. Figure 1–1 shows an air conditioning unit that is used as a computer room air conditioner.

FIGURE 1–1
Air conditioning unit used for cooling computer rooms
(Courtesy of Liebert Corp.)

Another example of a manufacturing process that uses an air conditioning system for temperature control is a printing plant. Temperature control and humidity control are important in a printing operation. The movement of paper through the printing machines and the flow of ink onto the paper are both affected by the conditions of the air in the room.

In many laboratory processes and in some manufacturing processes, the control of the environment in the rooms where the process occurs is critical. Examples are **clean rooms.** Air temperature, humidity, and cleanliness must all be maintained within very close tolerances in such rooms. Thus, cooling, humidification, and a high level of filtration are required. Figure 1–2 shows an air conditioning unit with a humidifier built into it for use in such an application.

Clean room

FIGURE 1–2
Self-contained air conditioning unit with humidifier built into
it (Courtesy of Liebert Corp.)

1.3

Description of Troubleshooting and Service

Troubleshooting

When an air conditioning or heat pump system is not
operating properly, the process of investigating the
problem and identifying the general area of the trouble
in the system is called **troubleshooting.** Troubleshooting
is performed by a technician who is familiar with the
normal operation of the system, who understands what
each of the individual parts in the system do, and who
understands the relationship of each part of the system
to all other parts.

When an air conditioning or heat pump system is
inoperative, a troubleshooter first finds out what the
general problem is. Then, he or she relates the general
problem to the performance of the major sections of the
system. Finally, the technician identifies the components of the system that are at fault. At that point, troubleshooting stops and service on the system begins.

Servicing an air conditioner or a heat pump system is the process of determining what part (or parts) in the system is at fault when the system does not perform as it should, and repairing or replacing that part. In most cases, an individual part is the cause of a problem, but in some cases, more than one part is faulty. Because the functions of the parts are interrelated, determining which part is malfunctioning is not always easy. Thus, a service technician must have knowledge of the parts and their function within a system. Service may be performed by the same person who troubleshoots a system, or it may be performed by an entirely different person.

Servicing

1.4

Systematic Approach

To properly troubleshoot or service an air conditioning or a heat pump system, a technician must understand the principles of refrigeration and know how the system is supposed to work. Also, the technician must know what tools, instruments, and supplies are needed to work on the equipment. In addition, he or she must be able to identify the major sections, the components, and the individual parts in a system. Finally, the technician must be able to check each part for proper operation. Each of these subjects is covered in this text. For the sake of clarity, the procedures followed in troubleshooting and in servicing an air conditioning system are treated as two separate subjects.

Throughout the text, emphasis is placed on the importance of using a systematic procedure for checking out a faulty system. A **systematic procedure** has the following steps:

Systematic procedure

1. A technician receives a complaint concerning the operation of a system. By a process of deduction, the technician isolates the major section of the system causing the problem.
2. The technician determines the components that are involved.
3. The technician determines the parts that are faulty.

The use of a systematic process saves time, energy, and money for both the technician and the owner of the system. Procedures for checking components and parts are described in this text for typical air conditioning and heat pump systems.

Troubleshooting charts
Service charts

Troubleshooting charts and service charts are tables or graphs that outline a systematic procedure for checking the sections, components, and parts of a system that is not working properly. They help technicians to identify the parts that are defective. Troubleshooting and service charts are discussed in this text, and complete charts are given in the appendix. These charts outline the systematic procedure in checking a system. Table 1–1 is an abbreviated portion of a troubleshooting chart for air conditioning. As indicated in the table, for a general complaint and a specific symptom, the chart lists possible causes of the problem and the major section of the system involved. While the procedures listed on these charts will not fit every conceivable situation, they do fit most complaints for a typical system.

1.5

Summary

Air conditioning and heat pump systems are used in many different applications. These applications include comfort cooling systems, production cooling systems, and critical-environment control systems. The continual

Complaint	Symptoms	Causes	Section, component, or part
No cooling	No air coming from supply registers	No electrical power to equipment	Electric power supply system
		Control system not functioning correctly	Operating controls
		Evaporator blower not operating	Safety controls
		Evaporator coil obstructed	Evaporator section
		Dampers closed	Air distribution system

TABLE 1–1 Portion of Typical Troubleshooting Chart for Air Conditioning

operation of a system in a comfort cooling application is desirable but not necessarily critical. However, the operation of a system in a process cooling or environmental control application may be absolutely necessary. So that systems do operate continually, or with as little delay as possible in case of a malfunction, technicians must be available to troubleshoot and service the equipment. Technicians can find and correct, quickly and efficiently, any problems that may occur in the system.

Troubleshooting is the process of analyzing system operation and determining the major sections and components of the inoperative system that are faulty. Service is the process of checking those components and/or parts of the system and repairing or replacing them if they are faulty.

The procedures to be followed in troubleshooting or servicing an air conditioning or heat pump system are outlined and described in troubleshooting and service charts. By using these charts, a technician can find problems and make corrections in a minimum amount of time. The two procedures, troubleshooting and service, are closely allied, and both may be performed by the same technician. However, they are actually two separate operations and are described as such in this text.

1.6

Questions

1. Define *troubleshooter*.
2. Define *service technician*.
3. Name three general types of air conditioning systems.
4. Explain the main use of comfort cooling systems.
5. A production cooling system is used for what purpose?
6. A critical-environment control system is used in what type of application?
7. Describe a systematic procedure for checking out a faulty system.
8. What is the purpose of a troubleshooting or service chart?

2

Tools, Instruments, and Supplies

2.1

Introduction

Troubleshooting and servicing an air conditioning or heat pump system require both thinking and doing. On a troubleshooting call, a technician must be able to analyze a system's performance in order to determine what is wrong. On a service call, a technician must be able to make repairs as needed. Troubleshooting requires knowledge of how the system operates. Servicing requires knowledge of how the components and parts work. Service also requires access to tools, instruments, and materials, with knowledge of how to use them in order to make the repairs.

The basic tools, instruments, and supplies used for service work are discussed in this chapter. The proper use for each is also described. The lists of tools, instruments, and supplies given here are not all-inclusive, but they can be used as a general guide to the items that a serviceperson will require for troubleshooting and service work.

2.2

Tools

Having the proper tools for working on equipment is part of the correct approach to troubleshooting and servicing an air conditioning or a heat pump system. Knowing how to use these tools is also necessary. Many of

the tools used by air conditioning and heat pump servicepeople are found in every shop and service truck.
Others are special tools required for specific operations.
The following descriptions cover most of the general
tools and many of the special ones.

Screwdriver

Wrench

Screwdrivers

A **screwdriver** is a tool used to install or remove screws
or to hold machine screws or bolts while nuts are placed
on them. Figure 2–1 shows an assortment of screwdrivers used for service work. A screwdriver has a metal
shaft with a handle on one end and a blade or other
special shape on the other end. The end of the blade is
shaped to fit a slot or recess in the head of a screw or a
bolt. In use, the shaped end of a screwdriver is placed
in the recess on the screw or bolt head, and the handle
is turned.

Different types of bolts and screws are used to hold
materials together. The types are characterized by the
shape of the recess in the head. The two most common
types are those with a plain slot and those with a cross-
shaped recess. Screwdrivers have specially shaped blades
to fit the type of screws they are used with. A straight,
or regular, screwdriver has a blade that is flattened on
the sides and is straight across the end. This blade fits
the slotted type of screw head. Screwdrivers that fit cross-
headed screws have a cross-shaped end. This end is
ground to fit the cross-shaped recess in the screw head.
Since screws and bolts come in many sizes, screwdrivers
also come in various sizes. A serviceperson should carry
a selection of sizes in each type.

Wrenches

A **wrench** is a tool used to hold a bolt while a nut is put
on it or to tighten the nut on a bolt. A wrench has a
head that fits on or around a nut, and it has a handle
that allows the user to turn the nut.

There are several types of wrenches. Each type is
designed and used for specific applications. Figure 2–2

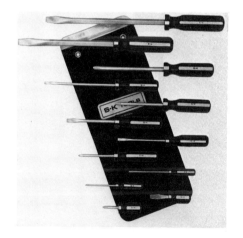

FIGURE 2–1
Assortment of screwdrivers used in
service work (Courtesy of S–K Tools)

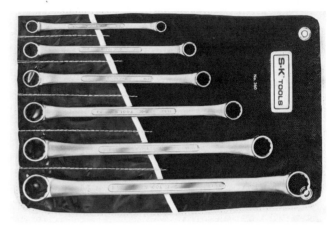

A. Open-End   B. Box-End

FIGURE 2–2   Wrenches used in servicing air conditioning equipment (Courtesy of S–K Tools)

shows two types of wrenches available, open-end (Figure 2–2A) and box-end (Figure 2–2B).

Other types of wrenches used in refrigeration work include adjustable, socket, and flare-nut wrenches. Also, nut drivers are used to fasten nuts. These tools are described in the following subsections.

Adjustable wrenches

Adjustable Wrenches.   With adjustable wrenches, one or both of the jaws in the head are movable. They can be adjusted to fit nuts within a certain range of sizes. Because of their movable head, they are not as strong as a wrench with a fixed head. So, they are used for general holding and tightening but not for heavy work.

Socket wrenches

Socket Wrenches.   A socket wrench is one of the most useful types of wrenches. The head, or socket, is separate from the handle, but it can be attached firmly to it. Thus, one handle can be used with a number of different-sized sockets. Figure 2–3 illustrates a set of sockets and a ratchet handle used with them. The sockets may also be used on different types of handles for different jobs.

The sides of the socket, or the part that fits over a nut, is one piece. When the wrench is placed on a nut,

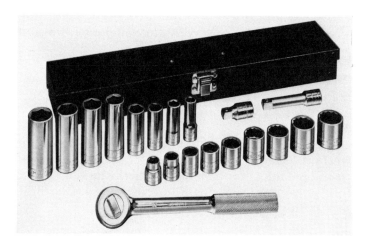

FIGURE 2–3
Set of socket wrenches and ratchet handle (Courtesy of S–K Tools)

the nut is completely enclosed by the walls of the socket. Thus, the socket wrench is a very strong tool. It is a good wrench to use when a nut needs to be very tight. Socket wrench sets come in various sizes and are available for all sizes of nuts.

Open-End Wrenches.   Open-end wrenches (Figure 2–2A) are the most common type of wrench. The head of the wrench is open on one side so that the wrench can be placed over a nut. The jaws on each side of the head are parallel and fit snugly on the sides of a nut.

Open-end wrenches

   Open-end wrenches come in a variety of sizes. Each wrench can be used on only one size of nut, so open-end wrenches come in sets to fit nuts through a range of sizes. Open-end wrenches are not very strong because of the open side. Hence, they are used for general holding and tightening, but they should not be used for heavy work.

Box-End Wrenches.   Box-end wrenches (Figure 2–2B) are similar to open-end wrenches, except that the opening in the head for the nut is completely enclosed. In use, the wrench is placed over the top of the nut. Box-end wrenches also come in various sizes, since each

FIGURE 2–4
Set of flare-nut wrenches used for refigerant tubing connections (Courtesy of S–K Tools)

wrench fits only one size of nut. Because of their closed construction, they are very strong. Therefore, they are used for heavy holding or tightening.

Flare-Nut Wrenches.   The flare-nut wrench is a special tool that combines the features of an open-end wrench and a box-end wrench. A set of flare-nut wrenches is shown in Figure 2–4.

The flare-nut wrench is used on flare nuts on refrigerant lines. The open end of the wrench is placed around the refrigerant line, and then the head is slipped down over the flare nut. The box shape of the head allows it to fit snugly over the nut.

Nut Drivers.   A set of nut drivers is a necessity for a service technician. A nut driver is a tool that looks like a box-end wrench on the end of a screwdriver shaft. The opening in the wrench head is in line with the shaft. The tool is used like a screwdriver, but the wrench head allows it to be used on nuts, bolts, or sheet metal screws.

Sheet metal screws are often used for fastening access panels on air conditioning or heat pump units. These screws have hexagon-shaped heads, which the nut drivers fit. Nut drivers are especially useful for rapid installation or removal of small nuts and screws.

Pliers

Pliers are hand-held tools that have jaws on one end. The jaws can be opened and closed by opening and closing the handles at the other end. They are designed for use with one hand. The jaws can be used to hold nuts or bolts, to twist wires together, or to hold a part when mechanical work is being done.

Pliers come in many different types, and each type is used primarily for one kind of work. An assortment of pliers is shown in Figure 2–5.

The diagonal cutting pliers shown in Figure 2–5A are used for general wire cutting and for stripping insulation from the end of wires for connection to terminals. Curved needle nose pliers, such as those shown in Figure 2–5B, are used for holding parts or fastenings

Box-end wrenches

Flare-nut wrenches

Nut drivers

Sheet metal screws

Pliers

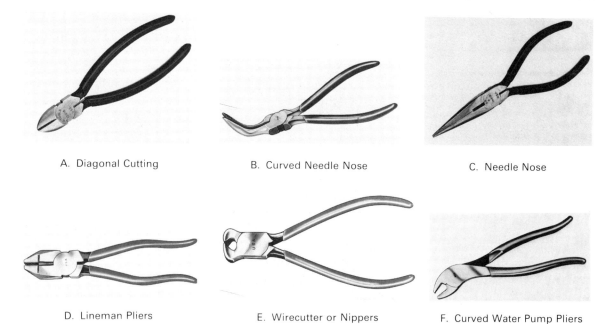

A. Diagonal Cutting   B. Curved Needle Nose   C. Needle Nose

D. Lineman Pliers   E. Wirecutter or Nippers   F. Curved Water Pump Pliers

FIGURE 2–5   Pliers used for service work (Courtesy of S–K Tools)

in small spaces where access is difficult with straight pliers. The needle nose pliers in Figure 2–5C are used for reaching into narrow spaces and holding parts or for attaching wires to terminals.

Figure 2–5D illustrates a pair of lineman pliers. These pliers are used for cutting wires, stripping insulation, and holding nuts or other wiring terminal fasteners. Wire cutters, or nippers, shown in Figure 2–5E, are used for cutting wires close to a fastener or for wire cutting in general. The curved water pump pliers shown in Figure 2–5F are used for holding parts and for tightening fasteners in places where straight pliers do not easily fit.

The pliers most commonly used in refrigeration work are water pump pliers, lineman pliers, slim-nose pliers, and vice grips. Each of these pliers is described in the following subsections.

Water Pump Pliers.   The most common type of pliers is the water pump pliers. In water pump pliers, the handles are fastened loosely at the pivot point by a pin

Water pump pliers

or a bolt. One of the handles has a slot that the pin goes through. This slot allows the size of the jaw opening to be adjusted somewhat. The inside of the jaws of the pliers are grooved for a better grip.

Water pump pliers are used primarily for holding or squeezing. They come in different sizes. The size used for a job is determined by the size of the material to be held.

Lineman pliers

Lineman Pliers.   Lineman pliers are a special adaptation of water pump pliers. These pliers have a wire cutter built in on one side of the jaws. They are usually used for cutting wires and for twisting wires together in electrical work.

Slim-nose pliers

Slim-Nose Pliers.   Slim-nose pliers are another adaptation of water pump pliers. The jaws of these pliers are extended and tapered. Slim-nose pliers are useful for reaching into restricted spaces or for working on small parts. They generally have a wire cutter on one side of the jaws and are often used for electrical work.

Vise grip pliers

Vise Grip Pliers.   Another special adaptation of water pump pliers is the vise grip pliers. For these pliers, the opening width of the jaws can be adjusted. Also, a cam arrangement on one handle allows the user to lock the pliers closed on the material it is holding. These pliers are especially useful for holding an object while it is being worked on.

Other Pliers.   There are many other types of pliers built for special jobs. Among these are wire strippers, angle pliers for reaching hard-to-get-at places, and spring-setting pliers. Each of these types of pliers has limited application, but each type is extremely useful when needed.

Miscellaneous Hand Tools

In addition to screwdrivers, wrenches, and pliers, a serviceperson should have some miscellaneous hand tools available for any job. Among these needed tools are hammers, chisels, punches, and tools used for applying special fasteners, such as rivet setters, electrical connection fasteners, piping or tubing cutters, and flaring tools.

While not strictly a hand tool, a small propane or butane torch for soldering or general heating is necessary. A portable oxyacetylene welding set with small tanks, such as the set shown in Figure 2–6, should be available for brazing, welding, and heavy heating.

2.3

Test Instruments

The first thing a serviceperson does on a typical service call is analyze the problem and determine the general section of the system that is causing the trouble. The next step is to check the components and parts in that section of the system. Either or both of these steps may require the use of test instruments. **Test instruments** are tools or equipment used for determining the performance of some part of a system.

Different test instruments are used for checking different parts, and test instruments fall into three general categories. The first category includes **general instruments**. General instruments are used for checking temperature, humidity, pressure, and air flow. The second category is **electric instruments**. Electric instruments are used to check the electrical characteristics of various parts and controls. The third category includes **mechanical instruments**. Mechanical instruments are used to check the mechanical performance of parts of the system. Each of these categories is discussed in the following subsections.

General Instruments

Thermometers. Thermometers are the most common and useful instruments used by air conditioning or heat pump servicepersons. A thermometer is an instrument for measuring temperature. Depending on the part of the system being tested, the temperature measured may be that of air or refrigerant.

Several different types of thermometers are used by service technicians. Among them is the common **stem thermometer**, which has a glass stem and bulb, as shown

Test instruments

General instruments

Electric instruments

Mechanical instruments

Thermometers

Stem thermometer

FIGURE 2–6
Portable oxyacetylene set for brazing or soldering (Courtesy of Union Carbide, Linde Div.)

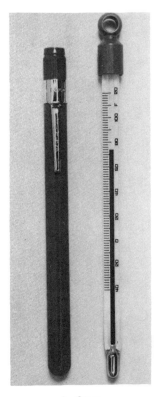

A. Stem

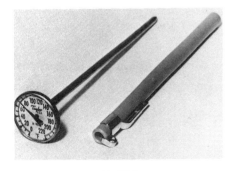

B. Dial

FIGURE 2–7
Thermometers used in service work
(Courtesy of Taylor Instruments/Sybron
Corp.)

in Figure 2–7A. The stem contains colored alcohol or mercury, and the temperature is shown on a calibrated scale on the stem. **Dial thermometers** are also commonly used in service work. A dial thermometer has a stem and a dial face at one end of the stem, as shown in Figure 2–7B. The stem is inserted into the material being tested, and the temperature is shown by the needle on the dial face.

Two common scales are used for indicating temperature: the Fahrenheit scale and the Celsius scale. The **Fahrenheit scale** is the one traditionally used in the United States. On this scale, water boils at 212 degrees and freezes at 32 degrees. The **Celsius scale** is used in the metric system and generally is used for scientific work. On this scale, water boils at 100 degrees and freezes at 0 degrees. The Celsius scale, along with other metric measurements, is being considered for worldwide use.

A serviceperson should have two stem thermometers available on any service call. The type with clips, so that it can be carried in a shirt pocket, is the most practical. In addition to pocket thermometers, a service technician should have two strap-on thermometers that can be attached to refrigerant lines in order to measure the refrigerant temperature. All service thermometers should have a wide-range scale so that they can be used to measure high or low temperatures.

Psychrometers.   A sling psychrometer is used to check the **humidity**, or the amount of moisture in the air. A sling psychrometer consists of two thermometers mounted side by side on a plate or a connecting device. One of the thermometers is a typical stem thermometer for measuring temperature. The other thermometer is similar, but it has a cotton wick on the bulb. The wick is fed by a small reservoir, which is filled with water.

The plate on which the thermometers is mounted has some type of handle that allows the user to rotate it rapidly through the air in order to get air to move past the bulbs. The dry thermometer indicates the sensible heat temperature, called the dry-bulb temperature. The thermometer with the wet wick indicates the amount of moisture in the air, and this reading is called the wet-bulb temperature. The wet-bulb reading will always be lower than the dry-bulb reading because the evaporation of the water in the wick cools the bulb of the wet-bulb thermometer. The difference between the two readings is an indication of the amount of moisture in the air. This difference is used with a slide rule or a psychrometric table to determine the relative humidity or other data concerning the conditions of the air.

Pressure Gauges. Pressure gauges are the second most useful instrument for the service technician. A typical pressure gauge for measuring liquid or gas pressure in a closed vessel is the **Bourdon tube gauge**. Figure 2–8 shows two typical Bourdon tube gauges. This gauge has a threaded port so that it can be attached to the part of the refrigeration system in which the pressure is to be read. The needle of the gauge moves in relation to the pressure, and the needle indicates the pressure on a calibrated scale on the gauge dial.

For air conditioning or heat pump service work, gauges are usually mounted in pairs on a **manifold**, or a pipe fitting with several outlets. One of the gauges, such as the gauge shown in Figure 2–8A, is called a compound gauge. A **compound gauge** measures pressure below atmospheric pressure, or vacuum, and pres-

Dial thermometers

Fahrenheit scale

Celsius scale

Psychrometers

Humidity

Pressure gauges

Bourdon tube gauge

Manifold

Compound gauge

A. Compound Gauge   B. High-Pressure Gauge

FIGURE 2–8
Pressure gauges for measuring refrigerant pressures (Courtesy of Robinair Division, Kent Moore Corp.)

High-pressure gauge

Manifold gauge set

Manometers

U tube manometer

Inclined manometer

sure above atmospheric pressure. This gauge is always used for measuring pressure in the low-pressure side of the system. The second gauge of the manifold is a high-pressure gauge, such as the one shown in Figure 2–8B. The **high-pressure gauge** does not have a scale for measuring vacuum pressure, but it does have a gauge for measuring pressure above atmospheric pressure. The high-pressure gauge is used for measuring pressure in the high-pressure side of the system.

The complete arrangement, called a **manifold gauge set**, is illustrated in Figure 2–9. A manifold gauge set has hose connections to each gauge, and valves allow the user to isolate the hose connections from each other and from a third hose connection. The hose connections can be used for connections to a refrigerant drum or a vacuum pump. The manifold gauge set allows a service technician to measure the pressure in both sides of a system simultaneously.

Manometers. A manometer measures low pressures. A monometer is most often used for measuring air pressure in some part of a duct system. The most common type of manometer used by a serviceperson is a **U tube manometer**. This manometer, shown in Figure 2–10, is a simple U-shaped tube of glass or plastic partially filled with water or other liquid with a known specific gravity.

In use, one side of the U tube manometer is connected to a duct with a rubber tube. The other side is left open to the atmosphere. Pressure of the air in the duct pushes the water down in the side of the U tube to which the hose is connected. This pressure in turn, causes the water level in the other tube to rise. The difference in the water level in the two legs of the U tube is an indication of the amount of pressure applied. The manometer is calibrated in inches, and the reading is in inches of water pressure or in inches water gauge.

Another common type of manometer is the **inclined manometer**. The inclined manometer is a U tube manometer with a wide spread between the legs and a long, sloping bottom section, as shown in Figure 2–11. The manometer is filled with liquid to a point near the upper part of the sloping bottom section. When it is used to measure pressure, a rubber tube is connected to the opening on the short leg of the U tube. The other end

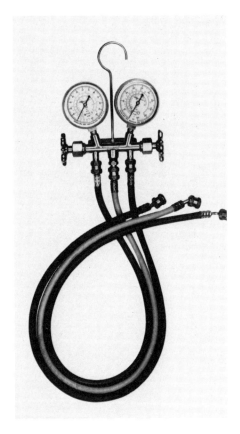

FIGURE 2–9
Manifold gauge set for checking pressures in both sides of a system at once (Courtesy of Robinair Division, Kent Moore Corp.)

of the rubber tube is connected to the part of the duct where a pressure reading is required. Pressure through the tube forces the liquid down and across the bottom section of the manometer. Since the bottom section is long in relation to its depth, the scale is long. Hence, these pressure readings are more accurate than readings taken with a conventional U tube manometer.

A more elaborate version of the U tube manometer, a **mercury manometer**, is used to measure a vacuum or very low pressures. A mercury manometer is a U-shaped glass tube that is open on one end to a threaded connection and is sealed on the other end. The threaded connection is hooked up to a refrigeration system when the system is being evacuated by a vacuum pump. The glass tube is filled with mercury, going from the sealed end, around the bottom, and up to a point on the side of the tube that is connected to the system. When a vacuum is drawn on the system, the mercury pulls away from the closed end of the tube. The greater the vacuum, the more the mercury moves. A calibrated scale runs along the tube on the closed side. The scale is calibrated in either inches or millimeters of mercury. The pressure, or vacuum, is then read in inches or millimeters.

Electric Instruments

Many electric controls are used on air conditioning equipment. And since many problems in the equipment are directly related to these controls, electric test instruments, or meters, are the most important instruments that the serviceperson carries. The basic electric meters used are voltmeters, ammeters, and ohmmeters. Figure 2–12 illustrates an assortment of these meters. Each type is discussed in the following subsections.

Voltmeters.   A voltmeter is an electric test meter that measures electric potential, or voltage, between two points in a circuit. The meter has two leads with probes on the ends. The probes are attached to, or touched to, the points in the circuit where a voltage reading is desired. A needle on the meter face indicates the voltage at that point in the circuit. Different meters have scales

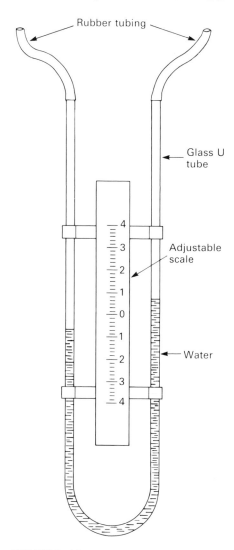

FIGURE 2–10
U tube manometer

Mercury manometer

Voltmeters

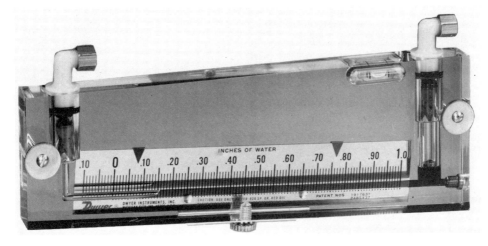

FIGURE 2–11
Inclined manometer for measuring low pressures (Courtesy of Dwyer Instruments, Inc.)

FIGURE 2–12
Electric test meters (Courtesy of Weston Instruments)

laid out in different ways, so the technician should read the owner's manual to become familiar with the particular meter being used.

Ammeters

Ammeters.   An ammeter measures the electric current in a circuit. Ammeters for measuring very small currents may be wired in series with a load. However, ammeters for measuring large currents are of the clamp-on type, and they should never be wired into the circuit. With the clamp-on ammeter, a set of jaws are clamped around one of the wires in an electric circuit. The current in the wire causes a flow of electricity in the jaws by induction. The meter indicates the amount of current, in amperes, by a needle on a calibrated face.

Ohmmeters.   An ohmmeter measures the resistance to electric current in a circuit. It can be used to check the resistance through motor windings, a relay coil, or any other part of a system. An ohmmeter has its own power source, usually dry-cell batteries, and it is never connected to an energized circuit. The ohmmeter has two leads with probes on the ends. When resistance is to be measured, the part of the system to be checked is isolated, or disconnected from the rest of the system. Then, the meter probes are connected to each side of the part of the circuit for which a reading is needed. The reading, in ohms, is shown on a calibrated dial by a needle or other indicator.

Multimeters.   Meters that are combinations of the meters just described are called multimeters. A common combination is a voltmeter and an ohmmeter. Figure 2–13 shows one type of multimeter. On a multimeter, the scales are arranged so that volts or ohms can be read from separate scales but are indicated by the same needle. A selection dial or switch is used to change from one type of reading to the other.

Other Meters.   Sometimes, other test meters for special applications may be needed by service technicians. These meters include wattmeters and capacitor checkers. Special meters should be available in the shop and need not be carried at all times by the technician.

Traditionally, all electric test meters had a needle that indicated the readings on a calibrated dial. But in many newer models, the reading is given by numerals on an electronic display. These meters are called **digital-readout meters**. They are very easy to read, and they reduce the possibility of errors that result from reading complicated scales. A digital meter is shown in Figure 2–14.

Ohmmeter

Multimeters

FIGURE 2–13
Multimeter for electrical testing
(Courtesy of Weston Instruments)

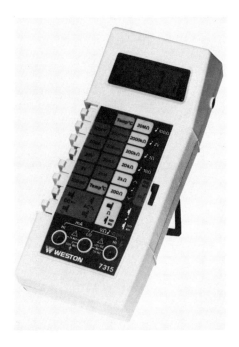

FIGURE 2–14
Digital meter for electrical testing
(Courtesy of Weston Instruments)

Digital-readout meters

Evacuation pumps

Charging cylinders

Leak detectors

Halide detector

While they are not actually meters, various jumper wires are often useful to a service technician. Jumper wires should always be made of heavier wire than the wire in the circuits they are used on. Also, they should have insulated clips on the ends for easy fastening to device terminals. Jumper sets with lights or buzzers are also useful for checking short circuits.

Mechanical Instruments

Mechanical instruments for troubleshooting and service work are used for checking the mechanical components and parts of a system. Mechanical instruments include evacuation pumps, charging cylinders, and leak detectors. Some of these instruments are used with other devices, but they are included in the following discussions because of their use as analytical tools.

Evacuation Pumps.   An evacuation pump, shown in Figure 2–15, is a vacuum pump that draws a very low pressure, or vacuum, on the refrigeration system. It is used to remove any water vapor in the system prior to charging with refrigerant. The pump should be capable of drawing close to 29.9 inches of mercury vacuum, a value that is close to a perfect vacuum. The pump is connected to the system through the hose connection on the center port of the manifold gauge set.

An accurate vacuum gauge should be used with an evacuation pump to indicate the value of the vacuum drawn on the system. A mercury manometer or an electronic gauge is used for this purpose.

Charging Cylinders.   When a technician must add refrigerant or charge a system with refrigerant, a charging cylinder is used. A charging cylinder is a glass cylinder that holds enough refrigerant, under pressure, to charge a system. The refrigerant in the cylinder remains in the liquid state because it is under pressure. As indicated in Figure 2–16, the cylinder is calibrated to show the amount of refrigerant it contains. It has a valve and a hose connection with which it is filled with refrigerant and connected to a system.

When the charging cylinder is in use, a hose is connected from the cylinder to the center connection on

FIGURE 2–15
Evacuation pump (Courtesy of Robinair Division, Kent Moore Corp.)

a gauge manifold. The gauge manifold is connected to the service ports on the system through the hoses on the pressure gauge connections. The refrigerant then flows out of the cylinder into the system through the gauge manifold. The calibrations on the cylinder allow the technician to measure the refrigerant as it flows out of the cylinder.

Leak Detectors.   Often, a troubleshooter or a service-person must look for leaks in the refrigerant lines or the coils of an air conditioning or heat pump system. To do so, he or she uses a leak detector.

Many methods can be used to locate leaks in a refrigeration system. The simplest method is to apply a soapsuds solution to the joints or suspected point of leaks. Bubbles in the soap solution will indicate the leaks. Mechanical or electronic leak detectors may also be used.

A good mechanical leak detector is a **halide detector**. As shown in Figure 2–17, a halide detector includes a propane torch that is used to heat a copper disk. A rubber tube is connected to the detector in such a way

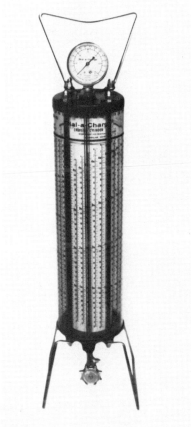

FIGURE 2–16
Charging cylinder for measuring refrigerants (Courtesy of Robinair Division, Kent Moore Corp.)

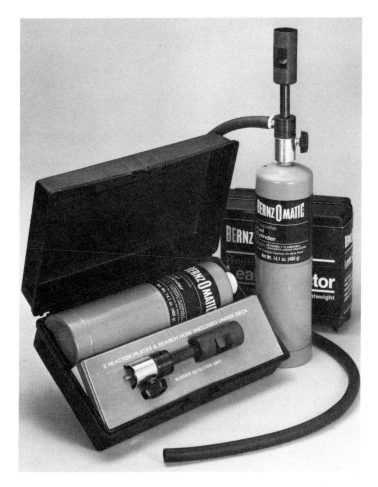

FIGURE 2–17
Halide leak detector (Courtesy of Bernz-O-Matic Corp.)

that air is constantly being drawn through it and mixed with the propane gas in the flame. A sample of air from around the refrigerant line or connection is drawn through the tube. If there is any refrigerant in the sample, the flame turns bright green. The chemicals in the refrigerant react with the copper to produce the green flame.

Electronic leak detector

An **electronic leak detector**, shown in Figure 2–18, is a device that analyzes a sample of air drawn from around the refrigerant lines or connections. It indicates

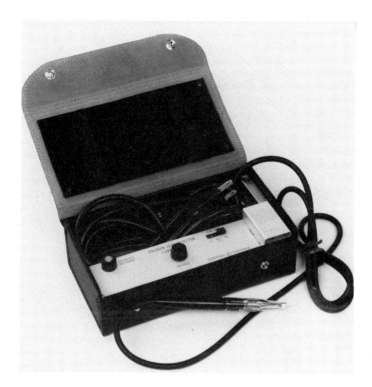

FIGURE 2–18
Electronic leak detector (Courtesy of General Electric,
Instrument Products Section)

whether there is any refrigerant in the sample. The sample is drawn through a tube that runs from the detector to the refrigerant line. The presence of refrigerant is indicated on a dial or by an audible signal from the detector. An electronic leak detector can be adjusted for sensitivity so that very small amounts of refrigerants can be detected.

Other Instruments.   In addition to the mechanical instruments just described, a serviceperson may need others on occasion. For example, pressure gauges for special applications, air flow meters, and tachometers for measuring the revolutions of motor or blower shafts may be needed. These special instruments can be kept in the service shop, available for use as needed.

2.4

Supplies and Equipment

In addition to test instruments and tools, the trouble-shooter or serviceperson needs supplies for the tools, instruments, and equipment. The serviceperson should also have information concerning the equipment that is worked on. This information is available from the equipment manufacturer in the form of engineering data sheets, installation manuals, and operation and service manuals. The supplies and equipment a technician needs are described in the following subsections.

Supplies for Test Instruments

Many of the test instruments used in troubleshooting and servicing require a fuel supply or other materials to make them operable. For example, fuel is needed for the torches used in soldering or welding. Soldered fittings are needed for connecting the refrigerant lines in a system. Welding rods, brazing rods, and silver solder, with appropriate fluxes and soldering pastes, are also needed. Finally, a supply of the common sizes of clean, dry refrigerant copper tubing is needed, along with a supply of fittings for the tubing.

Electrical Supplies

Since much of the troubleshooting and service work is performed on the electric components in an air conditioning or a heat pump system, a supply of the commonly used wire should be available. This wire includes insulated stranded wire for controls and insulated solid wire for power circuits. Various types and sizes of wire should be carried by the technician, according to the type of systems to be worked on. In addition, wire nuts for making connections should be carried. Mechanical connectors in a variety of types and sizes should also be carried, with a tool for attaching them to the wire ends. A technician should carry a stripping tool for removing the insulation from wire when connections are to be made. Finally, the technician should carry rubber or plastic grommets to protect the wire where it passes through knockout holes in metal panels.

Refrigerant Supplies

When the refrigeration system is serviced, the technician may need to replace some of the refrigerant in the system. To do so, the serviceperson needs a supply of the type of refrigerant used in the unit being serviced. Thus, service-sized tanks of the common refrigerants, as well as of nitrogen for protection against oxidation during soldering, should be carried by the serviceperson.

Figure 2–19 shows a selection of refrigerant service drums. Refrigerants are always identified on the tanks by an R number, and the tanks are also color-coded so that a service technician can quickly identify the proper refrigerant for each job. Tanks used for service work should be small enough to be carried conveniently in the service truck and on the job.

In some cases, a serviceperson will have to add oil to a compressor on the job. Thus, a service technician should carry a supply of high-grade refrigeration oil for this purpose. The oil should be one that is especially prepared and used for the lubrication of refrigeration compressors. The oil must be kept in a sealed can so that it will remain clean and dry.

Evacuation pumps also need lubricating oil for proper operation. Therefore, the serviceperson should carry a supply of oil suitable for use in the evacuation pump so that it will be available when needed on the job.

Equipment Information

Equipment information is needed by the serviceperson so that he or she can service and repair the system. This information is usually in the form of engineering data sheets, installation manuals, operation manuals, or service manuals. These items are furnished by the equipment manufacturer. Although all air conditioning and heat pump systems are similar in some respects, they are different in many aspects. So, knowledge concerning the specific unit being worked on is invaluable to the service technician.

FIGURE 2–19
Different sizes of refrigerant drums (Courtesy of E. I. duPont de Nemours and Co.)

2.5

Summary

This chapter discussed the tools, instruments, and supplies needed for troubleshooting or servicing an air conditioning or a heat pump system. The proper use of the devices was also covered. To troubleshoot or service an air conditioning or a heat pump system properly, a serviceperson must know how the system is supposed to work, know how it actually is working, understand the relationship between the various components and parts of the system, and, finally, know what tools, instruments, and materials are needed for the servicing and how to use each of them.

Tools are used to repair or replace a faulty part. In general, tools such as screwdrivers, pliers, and wrenches are the most commonly used. Special tools, such as wire strippers, cutters, and special wrenches, may be required at times for working on specific parts of a system.

Test instruments help the technician determine the performance of a system and its parts. General instruments check temperature, humidity, and pressure. Electric meters check the characteristics of the electric power supply and the electric controls. Mechanical instruments check the mechanical performance of parts of a system.

Since a service technician often makes repairs, either minor or major ones, the necessary supplies for the repairs must be available. Some of the supplies are related to the tools or the instruments used for service; an example is fuel for soldering torches. Other supplies, such as lubricating oil, are related to the operation of the equipment.

A properly equipped troubleshooter or serviceperson will have all the tools, instruments, and supplies required for a job whenever a service call is made. Having the correct equipment saves time and money for both the technician and the equipment owner.

2.6

Questions

1. A screwdriver is basically used for doing two things. Name them.

2. Describe the two types of screwdrivers that are defined by the types of screws they are used on.
3. A wrench is used to hold or tighten a _____ on a _____.
4. Name the five types of wrenches described in the chapter.
5. A wrench in which one or both of the jaws move is called an _____ wrench.
6. A wrench with individual heads that fit on a common handle and that entirely enclose a nut is called a _____ wrench.
7. A wrench with two parallel jaws and an opening on the side for placing it over a nut is called an _____ wrench.
8. A box-end wrench is similar to a _____, except that its head completely encloses a nut.
9. A _____ combines the features of an open-end wrench and a box-end wrench.
10. Pliers are hand-held tools used for holding nuts or _____ and for twisting _____ together.
11. Pliers in which the jaws can be clamped tightly on material are called _____.
12. Name three miscellaneous tools a serviceperson may need.
13. Test instruments are used to determine the _____ of some part of a system.
14. Name the two most common types of thermometers used by service personnel.
15. A psychrometer is used for measuring _____.
16. Pressure gauges for refrigeration work are mounted in pairs on an instrument called a _____.
17. Very low pressures are measured with a _____.
18. Name the three types of electric test meters commonly used by servicepeople.
19. Describe a multimeter.
20. What is an evacuation pump used for?
21. Describe a charging cylinder, and explain how it is used.
22. Name the three most common types of leak detectors used in service work.
23. Why are refrigerant cylinders color-coded?
24. Manufacturer's literature is a good source of information about a specific unit. Name three types of literature available.

CHAPTER

Hazards and Safety

3.1

Introduction

Hazard

A **hazard**, in relation to an air conditioning or a heat pump system, is any condition that creates a risk or a danger to persons working on or around the equipment. Hazards exist in air conditioning and heat pump troubleshooting and service work because some of the equipment worked on and some of the materials used have inherently hazardous properties. By understanding what these hazardous properties are, and by following safe procedures when working with the equipment, a troubleshooter or a serviceperson can prevent accidents.

Safety

Safety, as related to troubleshooting and servicing an air conditioning or a heat pump system, is the understanding and practice of safe procedures while working on the system. These safe procedures should minimize the possibility of accidental injury to the worker or to the equipment.

This chapter describes the hazards involved in the air conditioning field and the safety procedures technicians should follow the avoid accidents.

3.2

Hazards

Hazards for the air conditioning field fall into three categories: mechanical, electrical, and chemical. Safety procedures technicians should follow when working on the equipment belong to the same three categories. Each of the three categories of hazards is discussed in the following section, and safety procedures to avoid the hazards are described in the section after that.

Mechanical Hazards

Mechanical hazards exist because of the weight or mass of the parts of the system, because of the moving parts in the system, and in the use of the tools required for service work.

Mechanical hazards

A cluttered or dirty work area can contribute to the possibility of accidental injury to a worker. The potential for stumbling or falling over something or into some operating equipment is greatly increased when the work space is not clean.

When sheet metal is cut by a shears or when parts are machined in normal manufacturing operations, a sharp or rough edge is often left on the material. This sharp edge is a hazard to anyone working on the equipment since it can lead to cuts or abrasions when the hands, arms, or other parts of the body come into contact with the edge.

If a worker attempts to move a heavy part without mechanical aid, he or she may be injured. Moving parts in a system also pose some danger. For instance, a pulley-driven fan belt or a rotating wheel can cause serious injury to workers who get their fingers, hands, or arms in the way of the moving parts.

Some hazards result from the use of the tools required for service work. Wrenches, pliers, and screwdrivers can all inflict wounds if they slip during use and strike the worker using them. Or the tools may cause damage to components of the system.

Parts of a system that may be very hot always pose some threat to a technician because of the possibility of burns. A gas discharge line on a system or some parts of a heat pump may easily be hot enough to burn the skin during normal operation of the unit.

The following list presents a summary of some of the mechanical hazards encountered in typical air conditioning or heat pump workplaces.

- A dirty or cluttered work space.
- Oil or water on the floor of the work space.
- Poor lighting in the work area.
- Rough edges on the equipment or the covers.
- Sharp edges on the equipment or the parts.
- Sparks or hot particles thrown from the work material.
- High-temperature pipes or lines.
- Moving parts, pulleys, and wheels.
- High-pressure gases or liquids in parts of the system.
- Heavy weights to be lifted or moved.

Electrical Hazards

Electrical hazards

Electrical hazards are hazards related to the electric wiring and components in the air conditioning system. Electricity, by its very nature, is hazardous to a worker because it is not visible, and yet it can cause injury or death if it passes through a person's body. Many of the components and controls used in air conditioning systems are electric. Thus, there is electric power wiring to the unit and to various components within the unit.

Working on an electric appliance in a poorly lighted area is always hazardous. That is, the worker may not be able to see the wires or connections being worked on, and accidents may result.

Circuit wiring and connections are often exposed when access panels are removed for service. If a worker comes into contact with an exposed wire or connection, and if the worker is grounded in some way, an electric shock will occur. If the shock is severe enough, burns or even death may result.

The following list presents some of the typical electrical hazards that exist in air conditioning work.

- Poor lighting in the work space.
- Shocks from power circuits.
- Higher voltage in a circuit than was anticipated.
- Stored charge in a capacitor.
- Worn or frayed electric conductors.
- Water on the floor at the work location.
- A switch in a ground leg instead of a power leg.

Chemical Hazards

Chemical hazards exist because of the chemicals used in a system. Although few chemicals other than refrigerants are actually used in the operation of an air conditioning system, some are used for service procedures. Most of the chemicals used are not particularly dangerous if reasonable caution is taken in handling and working with them.

Good ventilation of the work area is always necessary when any chemical is being used for a service procedure. There must be adequate oxygen in the air to sustain life. Even though chemicals may not be toxic in themselves, when they are present in a gaseous state and replace the air in a work space, there may not be enough oxygen left for a person to breathe and survive.

Refrigerants used in most systems are chemical compounds called halocarbons. They are nonflammable, nontoxic, and nonpoisonous. And they are not injurious to health when breathed in small quantities for even extended periods of time. However, because they have a low boiling point, they can cause frostbite if they get on a person's skin. Also, they can cause considerable damage if they get into a worker's eyes.

Although most refrigerants are not hazardous, some refrigerants used in large systems *are* dangerous. The two most common are ammonia and sulfur dioxide. These refrigerants can cause permanent damage or death when breathed in even small quantities for a short period of time. Some chemical compounds make good refrigerants but are hazardous to use because they are extremely flammable. Among these chemicals are butane and propane, common fuel gases. Flammable gases should never be used as refrigerants.

Other chemicals found in air conditioning work include the various fuels used for welding and soldering torches and the chemical solutions used for cleaning coils. Most of the dangers inherent in fuels result from their flammability. Accidents can result if fuels are allowed to escape into the atmosphere or are transferred from one container to another in a closed space or in an area where there are open flames. Cleaning solutions used in air conditioning work are either strong acids or strong bases and are normally caustic. Thus, protection is necessary for a person's skin and eyes.

Chemical hazards

The following list presents the most common hazards related to the use of chemicals in air conditioning work.

- Poor ventilation in the work space.
- Poor lighting in the work space.
- High-pressure gases and liquids.
- Flammable materials.
- Caustic materials.
- Toxic materials.
- High temperatures.

Safety Precautions

To work safely on an air conditioning or a heat pump system, a troubleshooter or serviceperson must be aware of the hazards and must take care to avoid them. Some rather obvious observations can be made about working safely and carefully, but recognizing a hazard and avoiding any dangers presented by it will prevent accidents. Safety precautions for the mechanical, electrical, and chemical areas are presented in the following subsections.

Mechanical Safety

Whenever mechanical equipment is worked on, the work space should be cleared of any obstructions, and the area should be clean. Boxes or other material cluttering up the work area may cause a worker to fall or stumble into moving parts of the system. Also, water or oil on the floor can cause a person to slip and fall.

Proper lighting of the work area is necessary for mechanical work. A worker must be able to see the equipment and the parts when he or she is working on machinery with moving parts or using tools or instruments.

While sharp edges on metal parts cannot be avoided entirely, care must be taken when one is working with those parts in order to avoid injury.

Some repair processes can produce high temperatures. Thus, care must be taken when the worker is using tools such as grinders or torches so that the worker is not burned or a fire is not started on any material on the job.

Heavy parts should never be lifted by one worker alone. When any lifting is done, proper lifting procedures should be used. A worker should always keep the back straight when lifting and use the muscles in the legs to do the work. If the part to be lifted is very heavy, the worker should use a hoist or a lifting device, such as a forklift. When the worker is sliding a heavy part along the floor, he or she should not attempt to pull it but should get behind it and push it along. Again, if the part is too heavy to be moved readily, the worker should get mechanical help, such as a forklift or other moving device.

Moving parts and rotating machinery always pose a threat to a worker. The worker can easily forget about the part while working on a unit if it moves rapidly enough to become a mere blur. And moving wheels, belts, and pulleys often do just that. Constant vigilance must be practiced by the worker so that he or she does not inadvertently put a hand or arm into moving parts. Also, loose clothes or jewelry should never be worn by anyone doing service work. They pose a special threat when a person is working around moving machinery. Anything loose can be caught by moving parts, and the worker can thus be pulled into operating machinery.

Caution should always be exercised by anyone using tools. Tools can pinch, cut, or puncture a person's skin if they slip off the parts they are being used on. Tools should be kept in good repair and should be clean so that the possibility of their slipping is minimized. Also, they should be placed carefully on the part and held firmly in place while being used.

The following list summarizes the safety precautions that a technician should follow when working on the mechanical parts of a system.

- Always pull a wrench instead of pushing it in order to avoid damage to the nut and possible danger due to slipping.
- Use the proper type and size of tool for each job.

- Make sure that all mechanical connections on equipment are tight before operating it.
- Grind "mushroom heads" (flattened heads with rough edges) from chisels or hammer faces before using them.
- Wear goggles when drilling or sawing.
- Keep the floor of the work space clean and free of debris.
- Never allow spilled oil to remain on a floor.
- Always make sure that good lighting and ventilation are available in the work space.
- Use two wrenches when working on tubing fittings.
- Stay clear of moving pulleys and wheels.
- Do not wear loose clothing or jewelry when working on equipment with moving parts.
- Remove self-locking doors from any old refrigerator or freezer.
- Use proper hoisting and trucking equipment when lifting or moving heavy mechanical equipment.
- Keep a fire extinguisher available when using a torch.
- Do not let loose objects or tools fall into moving fans, pulleys, or belts.
- Fasten access doors and panels so that they cannot swing or fall over when equipment is worked on.
- Avoid touching hot gas discharge lines and other hot lines on equipment.
- Do not use an open torch to heat a closed drum.

Electrical Safety

Whenever a serviceperson works on the electric components in an air conditioning system, he or she must take care to see that there is proper lighting so that all parts and connections can be seen distinctly. The danger of shocks or damage to the instruments or parts is great if probes or jumpers are used incorrectly. If necessary, a work light on a drop cord should be used to bring light to the area before work is started on electric components.

The worker should always check the voltage of the system before starting any electrical work. A low-voltage system, such as a control system, is relatively safe to work on. But a high-voltage system, such as the power supply to a large compressor or blower motor, can be very dangerous. High-voltage systems may require special high-voltage probes on instruments to protect both the instrument and the operator from potential danger.

The electrical troubleshooter or serviceperson should not wear loose jewelry, such as rings, watches, or bracelets, that can make contact with the electric terminals or wires while the worker's hands are close to the electric parts. To avoid injury, a serviceperson should be insulated from the ground in case he or she does make accidental contact with a part of the electric system that is energized. To be safe, a serviceperson should never hold on to a pipe or other metal part with the free hand while working on electric parts. Also, the serviceperson should never stand in water or even on a wet spot on the floor when testing electric circuits. If necessary, the worker should stand on a dry board or other insulating material that will prevent grounding.

The following list is a partial list of safety precautions that a worker should observe when working on electric components in air conditioning systems.

- When working on electric circuits, determine whether the disconnect is open or closed.
- When working on a piece of equipment, always tag the disconnect to show that it shouldn't be opened.
- Make sure that the work space is well lighted.
- Never cut the grounding prong off of an electric plug for a tool or an appliance.
- Never use the green wire in an electric system for anything but a ground wire.
- Replace all ground straps and wires on equipment in order to protect the equipment and personnel in case of a short.
- Never connect AC equipment to DC power, or vice versa.
- Do not use test lights on solid-state circuits.
- Always be sure to be insulated from the ground when working on electric circuits.

- Always use insulated tools when working on electric circuits.
- Determine the voltage or amperage range of a circuit before using electric test meters.
- Always discharge a capacitor before touching the terminals.
- Replace worn or abraded electric wiring when it is found on equipment.
- Do not work on electric circuits if there is water on the floor of the work space.
- When working on electric circuits in a damp location, stand on insulating material.
- Do not wear loose jewelry when working on high-voltage circuits.
- Do not disconnect or bypass electric safety devices on equipment.

Chemical Safety

Proper lighting in the work space is just as important when one is working with chemicals as it is when one is working with mechanical or electric components. Positive identification of the materials being used and the procedures for their use is absolutely necessary when a worker is using chemicals. Poor lighting in the work space contributes to accidents and possible injury.

Because of the toxic nature of some chemicals, adequate ventilation is necessary when they are used. Even if a chemical is not toxic, when it is in the gaseous form, it may displace the oxygen in a work space and create a hazardous condition. Fresh air should circulate in the work space through doors or windows or from circulating fans.

Many of the chemicals used in air conditioning or heat pump systems or in service work are under pressure. The normal pressures found in operating systems are usually not dangerous. But a malfunction may produce abnormally high pressures, and this situation could be dangerous. Heat should never be applied with an open flame to a refrigerant drum or to any other drum containing material under pressure. If a drum or a tank must be heated, it should be placed in a pan of hot water.

Properly operating pressure gauges should be used on all pressurized drums so that the pressure of the material in the drum can be determined at any time.

A rupture in a refrigerant line or a drum can spray refrigerant over a serviceperson and cause serious frost burns. Gloves and protective clothing should be worn when a serviceperson is working with refrigerants. Goggles or a protective face mask should always be used whenever refrigerant is being transferred from a drum to a system or from one drum to another.

The hot gas discharge line—and, sometimes, the compressor, the compressor motor, or other motor in an operating system—operate at fairly high temperatures. These temperatures are high enough to burn a person's hands or arms if they come into contact with the parts. Thus, gloves and protective clothing should be worn when a person is working on any part of a system that has hot parts.

The soldering or welding of any part of an air conditioning system requires the use of torches or soldering irons. Torches are always a potential source of burns to the operator if they are not used properly. Gloves and protective clothing should always be worn when the person is working with torches of any kind.

The following list presents the safety precautions that a technician should follow when working with chemicals in air conditioning or heat pump systems.

- Always wear goggles when working with refrigerants or any caustic chemicals.
- Always wear gloves and protective clothing when working with chemicals.
- When using nitrogen while brazing or soldering tubing, always use a pressure gauge on the nitrogen drum.
- Never use oxygen to pressurize a system.
- Ventilate the work space when brazing or soldering.
- Prevent contact with the skin when using epoxy for leak repairs.
- Never face the glass covers on the gauges when opening valves on chemical drums.
- Always store all chemical gases, including refrigerants, in color-coded drums.

- Never refill any drum that is not marked for reuse.
- Never fill a refrigerant drum more than 85% full.
- Never warm a refrigerant drum with a torch. Use warm water.
- Open all valves slowly.
- Make sure that all refrigerant piping and fittings are kept clean and dry in storage.
- When working on an ammonia system or any system containing toxic gas, always ventilate the area in case of a leak.
- Do not store any refrigerant in a container not designed for the specific refrigerant.
- Check the R number of any refrigerant before using it.
- Do not use a torch or any open flame near a system with a large refrigerant leak.
- Always charge a refrigeration system into the low-pressure side.
- Handle oil from a burned-out system carefully; the oil can be highly acidic.
- Remove ice from refrigerator coils with hot water. Never use sharp objects.
- Never charge a system with more than 150 pounds per square inch when testing for leaks.
- Never heat nitrogen or carbon dioxide cylinders.
- Never use carbon tetrachloride for cleaning.
- Never open a refrigeration system until the pressure in the system has been reduced to atmospheric pressure.
- When purging refrigerant from a system, make sure that the refrigerant will not spray on any person or plant equipment.
- Make sure that both ends of a line that contains liquid refrigerant cannot be closed off tightly. Hydrostatic pressure can cause the line to burst.
- Store fuels for torches in a safe place, and make sure that fuel cylinders cannot fall.
- Do not tamper with any safety controls on a system.
- Always check the type of refrigerant used in a system before adding more.
- Keep the protective caps on all cylinders when they are not in use.
- Replace protective caps on service valves and expansion valves when they are not being used.

Summary

Some hazards are inherent in troubleshooting or servicing an air conditioning or heat pump system because of the nature of the equipment, tools, and materials used. Troubleshooting and service procedures must include consideration of the hazards so that any work performed will be done safely and in a way that avoids possible injury to the worker or damage to the equipment. Safe procedures will prevent accidents.

There are three main categories of hazards in air conditioning systems: mechanical, electrical, and chemical hazards. A serviceperson should be aware of these hazards and always be on guard against other hazards that may be unique to a specific system.

Every hazard is associated with safety procedures designed to avoid accident and injury. These safety procedures ensure that troubleshooting and servicing can be carried out without undue danger to the operator. Every troubleshooter and serviceperson should know and follow these safety procedures when working on an air conditioning system.

Questions

1. Define *hazard* as it relates to air conditioning or heat pump systems.
2. Define *safety* as it relates to service work.
3. Name three types of mechanical hazards associated with air conditioning or heat pump systems.
4. Name three types of electrical hazards associated with air conditioning or heat pump systems.
5. Name three types of chemical hazards associated with air conditioning or heat pump systems.
6. List three safety precautions that should be observed in relation to each of the following:
   a. Mechanical hazards
   b. Electrical hazards
   c. Chemical hazards
7. Who loses when an obvious hazard is overlooked and safety precautions are not taken?

CHAPTER

Troubleshooting and Service

4.1

Introduction

If an air conditioning or a heat pump system stops working the way it should, someone must check its operation and determine what is wrong with it. The person who does the checking must know how the system works when it is operating properly and should be able to recognize when it is not working properly. The checking of a malfunctioning system to determine what is wrong with it is called **troubleshooting**. As the name implies, a troubleshooter looks for the trouble when a system is not working the way it should.

To troubleshoot any system, a person must know how the system operates when it is working correctly, how the individual sections, components, and parts of the system work, and how the parts work together. An air conditioning or heat pump technician with a general knowledge of how a typical system works should be able to troubleshoot any system.

Service is the performance of the tasks required to fix any problems found in a heat pump or an air conditioning system during the troubleshooting procedure. It usually includes repair or replacement of components or parts that are faulty. A troubleshooter must have specific knowledge about the operation of the system, but a serviceperson must know how to make the repairs required.

Troubleshooting

Service

4.2

Troubleshooting

Troubleshooting is the process of checking an inoperative or improperly operating air conditioning or heat pump system to determine what is wrong with it. A troubleshooting call is usually initiated by a complaint that the system is not working properly. The job of the troubleshooter is to investigate the complaint, determine the symptoms of the system, and determine the causes of the symptoms. The following three steps are involved in troubleshooting.

1. A **complaint** is received from someone. The complaint is related to the performance of a system.

    Complaint

2. An analysis of the complaint will reveal some evidence of what is happening in the system—that is, will reveal the **symptoms**.

    Symptoms

3. An analysis of the symptoms will indicate what sections, components, or parts of the system are at fault—that is, will reveal the **causes** of the symptoms.

    Causes

When analyzing a complaint, a troubleshooter must get as much information as possible about how the system is operating. This information includes a description of how the system is operating presently, as well as how it worked before any malfunction. It is usually acquired by questioning the person making the complaint.

An analysis of the symptoms will usually point to one of the main components of the system. Table 4–1 illustrates how various symptoms indicate problems with specific major components. For example, according to the table, a symptom of the air being too warm as it leaves the registers indicates one of the following:

*   A problem with the compressor,
*   A problem with the expansion device,
*   A problem with the condensing unit,
*   A problem in the air distribution system.

Symptom	Major component
No air coming from the supply registers	Evaporator blower
	Balancing dampers
	Fire dampers
War air coming from the supply registers	Compressor
	Expansion device
	Condensing unit
Room air temperature higher than set point temperature of the thermostat setting	Compressor
	Expansion device
	Condensing unit
Air too warm from the supply registers	Compressor
	Expansion device
	Condensing unit
	Air distribution system
Room air temperature lower than the thermostat setting	Equipment undersized
	Air distribution system
Cool drafts in the area	Air distribution system
Popping or ticking noises from registers	Air distribution system
Scraping or squeaking noises from registers	Air distribution system
Noisy outdoor unit	Compressor
	Condenser blower
Vibration or rattling in building	Refrigerant lines
Unit runs constantly	Controls
Unit needs frequent service	General equipment

TABLE 4–1   Major Components of System Related to Symptoms

The final results of the process include the identification of the major section (or sections) of the system that is involved in the problem and identification of any individual parts that are faulty. The steps in the troubleshooting process should be followed in an orderly and systematic manner so that the faulty parts of the system can be located as quickly and efficiently as possible.

To troubleshoot either an air conditioning or a heat pump system, a technician must be familiar with how each system operates when it is functioning correctly as well as how it is operating when the troubleshooting call is made. The troubleshooter must be familiar with the major sections of the system, the main components, and the typical parts. Some special tools and instruments may be required for checking out various parts of the system, and the troubleshooter must be familiar with the use of these tools.

4.3

Service

Since troubleshooting is primarily used to *identify* problems in an air conditioning or a heat pump system, a follow-up call to a troubleshooting call must be made in order to repair or replace the faulty parts. This call is the function of a serviceperson. In some cases, service may be performed by the same person who does the troubleshooting. But the two tasks are completely separate and are often performed by two different people.

After a troubleshooter analyzes the complaint, identifies the symptoms, and finds the causes of a problem, either the troubleshooter or the serviceperson must identify the faulty part (or parts). After the faulty parts have been found, they must be repaired or replaced. Repair and replacement of faulty parts are the jobs of a serviceperson.

The three steps shown earlier for troubleshooting are paralleled in the service process. That is, the complaint is compared to the major section of the system involved in the problem. The symptoms are related to the main components within the section. And the causes are related to individual parts. Thus the three steps in the service procedure are related to troubleshooting as follows:

1. The **major section** in which the problem lies is usually indicated directly by the type of complaint received.   Major section
2. The main **components** of the system within the major section are often directly related to the symptoms.   Components
3. The faulty **parts** can normally be isolated by reference to the causes, as identified in the troubleshooting procedure.   Parts

A comparative chart showing how the trouble-shooting steps carry over into the service steps is shown in Table 4–2. The example shown is for a complaint of "no cooling." A problem of "no air out of the supply registers" is indicated. The most likely major section and components of the system that may be involved in the problem are listed next. And finally, the parts to be checked when searching for the problem are listed.

A serviceperson must be familiar with the use of tools for general mechanical repairs. The serviceperson also must be familiar with, and know how to use, special tools and instruments for checking the refrigeration systems in the equipment.

After the faulty parts of an inoperative system are repaired or replaced, the equipment must be run through a normal operation cycle to make sure that the problem has been solved. Thus, the service technician must understand the operational cycle of the system. Tests should be made to determine that all of the components and parts are functioning as they should and to ensure that further follow-up calls will not be necessary.

Summary

Troubleshooting is the identification of any problem in an air conditioning or a heat pump system when it is

Complaint	Problem	Major section	Components	Parts	Reference
No cooling	No air coming from supply registers	Electric power to unit	Main power supply	Branch; disconnect	Chap. 9
				Wiring	Chap. 9
				Fusing or breakers	Chap. 9
				Unit; disconnect	Chap. 9
				Wiring	Chap. 9
				Fusing or breakers	Chap. 9

TABLE 4–2   Portion of a Service Chart

not operating the way it should. The troubleshooting process includes analyzing complaints, identifying the major section of the system in which the fault lies, and identifying the faulty part (or parts) in the section. A troubleshooter must be familiar with special tools and instruments that are used for troubleshooting and must know how to use them.

Service is the repair or replacement of faulty parts in a heat pump or an air conditioning system that is not working as it should. A serviceperson should be capable of using all tools and instruments required for service work. The serviceperson should also know how individual parts, individual components, and the entire system work in normal operation so that the operation can be checked out after repairs are made.

4.5

Questions

1. Describe the process of troubleshooting.
2. What are the final results of the troubleshooting procedure?
3. List the two things that are necessary for a person to be able to successfully troubleshoot a system.
4. What is the necessary follow-up to the trouble-shooting process?
5. Describe the process of service.
6. What are the necessary follow-up steps related to the service procedure?

Refrigeration System

5.1

Introduction

An air conditioning or a heat pump system is a refrigeration system used to cool or heat air. To understand the operation of an air conditioning or a heat pump system and to be able to service either of them, a technician must understand the refrigeration system.

This chapter explains the operation of a refrigeration system and describes the major sections of the system and the major components within the major sections. The relationship between the refrigeration system and the air conditioning or the heat pump system is also explained.

5.2

Principles of Refrigeration

Refrigeration system

A **refrigeration system** is a mechanical system that moves heat from one place to another. The heat is carried by a refrigerant that is circulated through a closed system of piping and coils. The refrigerant picks up heat from a place where it is not wanted and rejects it to a place where it is not objectionable.

A good example of a refrigeration system is a domestic refrigerator. In a refrigerator, the refrigerant circulates through the coils inside the refrigerator, where it picks up heat from the food, as indicated in the schematic drawing of Figure 5–1. The refrigerant is then drawn out of the coil by a compressor and is circulated

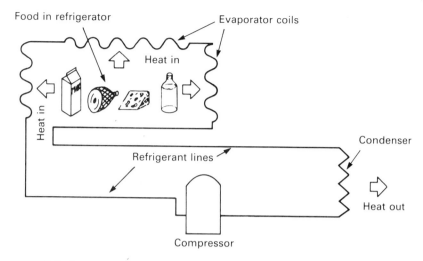

Food in refrigerator

Evaporator coils

Heat in

Heat in

Condenser

Refrigerant lines

Heat out

Compressor

FIGURE 5–1
Schematic drawing of refrigeration system in a domestic refrigerator

to another coil outside the refrigerator. This coil is usually located on the back of the refrigerator. The heat that is picked up from the food inside the refrigerator is given up to the room air from the coil on the back.

The refrigeration system is made up of two main sections, which include various components, individual parts, and the refrigerant lines. The two main sections are the **evaporator section**, which includes the expansion device, and the **condensing section**, which includes the compressor. The evaporator section is called the **low-pressure side** of the system, because the refrigerant is at a relatively low pressure in this part of the system. The condensing section is called the **high-pressure side** of the system, since the refrigerant is at a relatively high pressure in this part of the system.

The major components in the evaporator, or low-pressure side of the system, are the expansion device, the evaporator coil, and the refrigerant line that runs from the evaporator coil to the compressor. This line is called the **suction line**.

The major components in the condensing section, or high-pressure side of the system, are the compressor, the condensing coil, and the two refrigerant lines that run from the compressor to the condensing coil and from the condensing coil to the expansion device. These lines

Evaporator section
Condensing section

Low-pressure side

High-pressure side

Suction line

Hot gas discharge line
Liquid line

Compressor

Expansion device

are called the **hot gas discharge line** and the **liquid line**, respectively.

The refrigerant that circulates through the system is a chemical compound that boils at a relatively low temperature. The boiling point temperature of the refrigerant, at the pressure maintained in the evaporator coil, is lower than the temperature of the air around the evaporator coils. The condensing temperature of the refrigerant in the condenser coil is higher than the temperature of the air around the condenser coil. The boiling point temperatures in the two parts of the system are controlled by the pressures in those parts. Different refrigerants have different boiling point temperatures at different pressures. A particular refrigerant is chosen for each system according to the temperatures desired in the evaporator and the condenser sections of the systems.

The **compressor** creates the high pressure on the condensing side of the system and circulates the refrigerant. The **expansion device** causes a pressure drop in the refrigerant and maintains the low pressure in the evaporator side of the system.

5.3

Major Sections of a Refrigeration System

The two major sections in a refrigeration system are the low-pressure side of the system and the high-pressure side of the system. Heat is removed from the material to be cooled in the low-pressure side of the system, and that heat is expelled in the high-pressure side of the system. The two sections are illustrated in Figure 5–2. The refrigerant circulating through the system carries the heat. Various components and parts within the sections control the flow of the refrigerant.

Low-Pressure Side of the System

The main components in the low-pressure side of the system are the expansion device, the evaporator section, and the refrigerant line that connects the evaporator to the compressor. Figure 5–3 illustrates this part of the system.

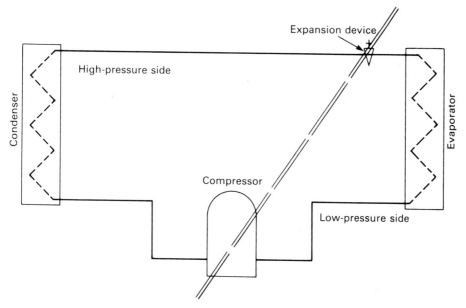

FIGURE 5–2
Two major sections of refrigeration system

The expansion device causes a drop in pressure in the refrigerant as it enters the evaporator section. This pressure drop is accompanied by a temperature drop. The low-temperature refrigerant then flows through the evaporator coil, where heat is absorbed from the cooling medium. The refrigerant is then drawn from the evaporator by the compressor through the refrigerant line that connects the two. This line is the suction line.

High-Pressure Side of the System

The main components in the high-pressure side of the system are the compressor, the hot gas discharge line, the condensing section, and the liquid line. Figure 5–4 illustrates this part of the system.

The refrigerant pressure is raised in the compressor, and its temperature is consequently increased. The high-pressure, high-temperature refrigerant leaves the compressor through the hot gas discharge line and flows to the condensing section of the system. As the refrigerant flows through the coil in the condensing section, it gives up the heat that it received in the evaporator and compressor. The refrigerant condenses to a liquid as it

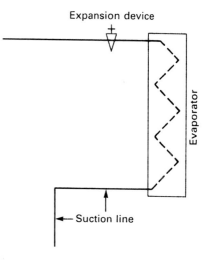

FIGURE 5–3
Components in low-pressure side of refrigeration system

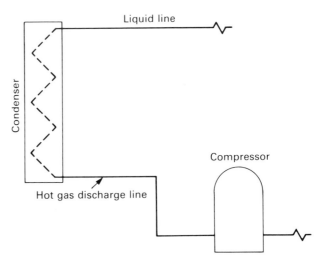

FIGURE 5–4
Components in high-pressure side of refrigeration system

gives up the heat. The liquid refrigerant flows from the condensing coil to the expansion device through the liquid line. The refrigerant is then ready to start another cycle through the system.

The refrigerant lines that connect the various parts of the system are necessary for the operation of the system. But they also serve as connections between the components and parts.

Controls to turn the system on and off and to regulate the cooling capacity of the system are important parts of the total system. Safety controls are also important. The function of the controls in a refrigeration system is not covered in this text, but air conditioning and heat pump system controls are discussed in Chapter 12.

5.4

Refrigerants

Refrigerant

By definition, a **refrigerant** is a fluid used for heat transfer in a refrigeration system. It absorbs heat at a low pressure and temperature and rejects heat at a higher pressure and temperature. The changes in pressure and temperature are accompanied by a change of state.

Various chemical compounds can be used as refrigerants, and many different ones have been employed since refrigeration systems were first built. This section examines the historical background of the use of refrigerants and the pressure/temperature relationships of refrigerants.

Historical Background

Probably, the earliest use of a refrigerant for cooling was the use of ice for cooling food and drinks. In this example, water is used as the refrigerant (ice is water in the solid state). As ice melts, or changes state from solid to liquid, it picks up heat from the water and air around it and cools them. Water is an excellent refrigerant and is still used today in many applications.

During the late 1800s and early 1900s, mechanical refrigeration systems using air or chemicals as refrigerants became fairly common. The chemicals used were ammonia, carbon dioxide, sulfur dioxide, methyl chloride, and, in some cases, various hydrocarbons. While these chemicals are excellent refrigerants, they have some objectionable properties, such as toxicity or flammability. Hence, they are used less frequently today.

In the late 1920s, a new family of refrigerants was developed. These refrigerants are chemical compounds composed of fluorine and carbon and are called **fluorocarbons**. They can carry reasonable amounts of heat, and they are also nontoxic, nonflammable, and relatively safe to use. These refrigerants were developed by General Motors Company, who later manufactured them jointly with duPont. Later still, duPont acquired exclusive manufacturing rights, and the refrigerants were manufactured by the Freon division of that company. Refrigerants are still often called by the general name Freon. They are also given an R number that identifies the particular chemical composition of each type.

Fluorocarbons

Pressure/Temperature Relationships

The **boiling point temperature**, or the temperature at which a material will vaporize or condense, of all ma-

Boiling point temperature

terials is a function of the pressure of the material. Thus, if the pressure changes, the temperature at which that material will vaporize or condense will also change. For example, water boils at 212 degrees Fahrenheit at normal atmospheric pressure of 29.96 pounds per square inch. If the pressure increases, the temperature at which the water boils increases. And if the pressure decreases, the temperature of the boiling point decreases. Refrigerants follow this same general law. Thus, controlling the pressure in the two main sections of a refrigeration system provides control of the boiling point temperature as well. In turn, the operating conditions of the system can be controlled.

The boiling point temperatures of refrigerants at any given pressure are shown on pressure/temperature charts or graphs. One such chart is shown in Table 5–1 for some common refrigerants. Notice that the refrigerants are identified by numbers. To use the chart, find the refrigerant you want to use by number. This number is shown at the top of the chart. Locate the pressure directly under the refrigerant in the column, and find the boiling point temperature of that refrigerant at that pressure in the left-hand column of the chart.

TEMP F	"Freon" Refrigerants							
	11	12	13	22	113	114	500	502
−50	28.9	15.4	57.0	6.2	—	27.1	13.1	0.0
−48	28.8	14.6	60.0	4.8	—	26.9	12.1	0.7
−46	28.7	13.8	63.1	3.4	—	26.7	11.1	1.5
−44	28.6	12.9	66.2	2.0	—	26.5	10.1	2.3
−42	28.5	11.9	69.4	0.5	—	26.3	9.0	3.2
−40	28.4	11.0	72.7	0.5	—	26.0	7.9	4.1
−38	28.3	10.0	76.1	1.3	—	25.8	6.7	5.1
−36	28.2	8.9	79.7	2.2	—	25.5	5.4	6.0
−34	28.1	7.8	83.3	3.0	—	25.2	4.2	7.0
−32	27.9	6.7	87.0	3.9	—	25.0	2.8	8.1
−30	27.8	5.5	90.9	4.9	29.3	24.6	1.4	9.2
−28	27.7	4.3	94.9	5.9	29.3	24.3	0.0	10.3
−26	27.5	3.0	98.9	6.9	29.2	24.0	0.8	11.5
−24	27.4	1.6	103.0	7.9	29.2	23.6	1.5	12.7
−22	27.2	0.3	107.3	9.0	29.1	23.2	2.3	14.0

TABLE 5–1   Pressure-temperature relationships

TEMP F	"Freon" Refrigerants							
	11	12	13	22	113	114	500	502
−20	27.0	0.6	111.7	10.1	29.1	22.9	3.1	15.3
−18	26.8	1.3	116.2	11.3	29.0	22.4	4.0	16.7
−16	26.6	2.1	120.8	12.5	28.9	22.0	4.9	18.1
−14	26.4	2.8	125.5	13.8	28.9	21.6	5.8	19.5
−12	26.2	3.7	130.4	15.1	28.8	21.1	6.8	21.0
−10	26.0	4.5	135.4	16.5	28.7	20.6	7.8	22.6
− 8	25.8	5.4	140.5	17.9	28.6	20.1	8.8	24.2
− 6	25.5	6.3	145.8	19.3	28.5	19.6	9.9	25.8
− 4	25.3	7.2	151.1	20.8	28.4	19.0	11.0	27.5
− 2	25.0	8.2	156.5	22.4	28.3	18.4	12.1	29.3
0	24.7	9.2	162.1	24.0	28.2	17.8	13.3	31.1
2	24.4	10.2	167.8	25.6	28.1	17.2	14.5	32.9
4	24.1	11.2	173.7	27.3	28.0	16.5	15.7	34.8
6	23.8	12.3	179.7	29.1	27.9	15.8	17.0	36.9
8	23.4	13.5	185.8	30.9	27.7	15.1	18.4	38.9
10	23.1	14.6	192.1	32.8	27.6	14.4	19.8	41.0
12	22.7	15.8	198.5	34.7	27.5	13.6	21.2	43.2
14	22.3	17.1	205.7	36.7	27.3	12.8	22.7	45.4
16	21.9	18.4	211.9	38.7	27.1	12.0	24.2	47.7
18	21.5	19.7	218.7	40.9	27.0	11.1	25.7	50.0
20	21.1	21.0	225.7	43.0	26.8	10.2	27.3	52.5
22	20.6	22.4	232.9	45.3	26.6	9.3	29.0	54.9
24	20.1	23.9	240.2	47.6	26.4	8.3	30.7	57.5
26	19.7	25.4	247.7	49.9	26.2	7.3	32.5	60.1
28	19.1	26.9	255.4	52.4	26.0	6.3	34.3	62.8
30	18.6	28.5	263.2	54.9	25.8	5.2	36.1	65.6
32	18.1	30.1	271.2	57.5	25.6	4.1	38.0	68.4
34	17.5	31.7	279.4	60.1	25.3	2.9	40.0	71.3
36	16.9	33.4	287.7	62.8	25.1	1.7	42.0	74.3
38	16.3	35.2	296.2	65.6	24.8	0.6	44.1	77.4
40	15.6	37.0	304.9	68.5	24.5	0.4	46.2	80.5
42	15.0	38.8	313.9	71.5	24.2	1.0	46.4	83.8
44	14.3	40.7	322.9	74.5	23.9	1.7	50.7	87.0
46	13.6	42.7	332.2	77.6	23.6	2.4	53.0	90.4
48	12.8	44.7	341.5	80.8	23.3	3.1	55.4	93.9
50	12.0	46.7	351.2	84.0	22.9	3.8	57.8	97.4
52	11.2	48.8	360.9	87.4	22.6	4.6	60.3	101.1
54	10.4	51.0	371.0	90.8	22.2	5.4	62.9	104.8
56	9.6	53.2	381.2	94.3	21.8	6.2	65.5	108.6
58	8.7	55.4	391.6	97.9	21.4	7.0	68.2	112.4
60	7.8	57.7	402.3	101.6	21.0	7.9	71.0	116.4
62	6.8	60.1	413.3	105.4	20.6	8.8	73.8	120.5

TABLE 5–1   (continued)

TEMP	"Freon" Refrigerants							
F	11	12	13	22	113	114	500	502
64	5.9	62.5	424.3	109.3	20.1	9.7	76.7	124.6
66	4.9	65.0	435.4	113.2	19.7	10.6	79.7	128.9
68	3.8	67.6	446.9	117.3	19.2	11.6	82.7	133.2
70	2.8	70.2	458.7	121.4	18.7	12.6	85.8	137.6
72	1.6	72.9	470.6	125.7	18.2	13.6	89.0	142.2
74	0.5	75.6	482.9	130.0	17.6	14.6	92.3	146.8
76	0.3	78.4	495.3	134.5	17.1	15.7	95.6	151.5
78	0.9	81.3	508.0	139.0	16.5	16.8	99.0	156.3
80	1.5	84.2	520.8	143.6	15.9	18.0	102.5	161.2
82	2.2	87.2	534.0	148.4	15.3	19.1	106.1	166.2
84	2.8	90.2	—	153.2	14.6	20.3	109.7	171.4
86	3.5	93.3	—	158.2	13.9	21.6	113.4	176.6
88	4.2	96.5	—	163.2	13.2	22.8	117.3	181.9
90	4.9	99.8	—	168.4	12.5	24.1	121.2	187.4
92	5.6	103.1	—	173.7	11.8	25.5	125.1	192.9
94	6.4	106.5	—	179.1	11.0	26.8	129.2	198.6
96	7.1	110.0	—	184.6	10.2	28.2	133.3	204.3
98	7.9	113.5	—	190.2	9.4	29.7	137.6	210.2
100	8.8	117.2	—	195.9	8.6	31.2	141.9	216.2
102	9.6	120.9	—	201.8	7.7	32.7	146.3	222.3
104	10.5	124.6	—	207.7	6.8	34.2	150.9	228.5
106	11.3	128.5	—	213.8	5.9	35.8	155.4	234.9
108	12.3	132.4	—	220.0	4.9	37.4	160.1	241.3
110	13.1	136.4	—	226.4	4.0	39.1	164.9	247.9
112	14.2	140.5	—	232.8	3.0	40.8	169.8	254.6
114	15.1	144.7	—	239.4	1.9	42.5	174.8	261.5
116	16.1	148.9	—	246.1	0.8	44.3	179.9	268.4
118	17.2	153.2	—	252.9	0.1	46.1	185.0	275.5
120	18.2	157.7	—	259.9	0.7	48.0	190.3	282.7
122	19.3	162.2	—	267.0	1.3	49.9	195.7	290.1
124	20.5	166.7	—	274.3	1.9	51.9	201.2	297.6
126	21.6	171.4	—	281.6	2.5	53.8	206.7	305.2
128	22.8	176.2	—	289.1	3.1	55.9	212.4	312.9
130	24.0	181.0	—	296.8	3.7	58.0	218.2	320.8
132	25.2	185.9	—	304.6	4.4	60.1	224.1	328.9
134	26.5	191.0	—	312.5	5.1	62.3	230.1	337.1
136	27.8	196.1	—	320.6	5.8	64.5	236.3	345.4
138	29.1	201.3	—	328.9	6.5	66.7	242.5	353.9

Courtesy of E. I. duPont de Nemours

TABLE 5–1   (continued)

5.5

Operation of a Typical Refrigeration System

Refrigerant R–12, which is a designation for a particular refrigerant, is often used in typical refrigeration systems. As the refrigerant flows through the evaporator section of the system, it is at a pressure of about 21 pounds per square inch. The boiling point temperature of R–12 at this pressure is approximately 20 degrees Fahrenheit. Thus, the temperature of the refrigerant in the coil is 20 degrees Fahrenheit. If the temperature of the refrigerator box gets any warmer than 20 degrees when food is placed in the box, the heat from the food is absorbed by the refrigerant. As heat is absorbed by the refrigerant, the refrigerant boils or vaporizes, and the heat is removed as the vapor is moved from the coil by the compressor.

Figure 5–5 shows the relative temperatures of the refrigerant in the evaporator coils and of the air inside the refrigerator box in a typical situation. Heat is absorbed from the food and the air in the refrigerator box because of the temperature difference between the air and the refrigerant. This heat transfer causes the refrigerant to boil, or change from the liquid state to the vapor state.

The refrigerant vapor is drawn from the evaporator by the compressor. The pressure of the refrigerant is raised in the compressor to about 125 pounds per square inch. The refrigerant temperature is also raised because of the heat added by the heat of compression and because of the cooling of the compressor motor windings in certain types of compressors. The saturated temperature of the refrigerant is about 105 degrees Fahrenheit, but the actual temperature in the hot gas discharge line is much higher. The actual temperature is higher because the refrigerant is superheated. The **saturated temperature** is the temperature at which the change of state takes place, and **superheat** is the heat added to the refrigerant after it has completely turned to vapor.

The hot refrigerant gas leaves the compressor through the hot gas discharge line and goes through the condensing coil. The temperature of the refrigerant in

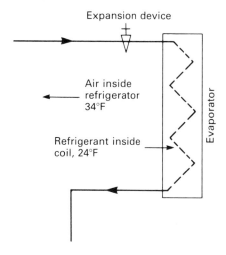

FIGURE 5–5
Relative temperatures of refrigerant in evaporator coils and of air in refrigerator box

Saturated temperature

Superheat

the coil is 105 degrees plus any superheat. The temperature of the air surrounding the coil, which is room temperature, is lower than the temperature of the refrigerant in the coil, so heat flows from the refrigerant to the air. Figure 5–6 shows the relative temperatures of the refrigerant in the condensing coil and of the air surrounding the coil in a typical situation.

The amount of heat given up by the refrigerant in the condenser is the same as the heat picked up in the evaporator and the compressor. As heat is given up by the refrigerant in the condenser coil, the refrigerant changes state from vapor to liquid. The refrigerant leaves the condensing section at the same pressure as it entered, but it is now a high-pressure liquid. It then flows to the expansion device through the liquid line.

This high-pressure liquid flows through the expansion device, where the restriction within the device causes a pressure drop. In this example, the pressure drops to 21 pounds per square inch, which is the pressure in the low-pressure side of the system. The reduction in pressure allows a certain amount of the refrigerant to flash into gas, causing a drop in temperature. Thus, the refrigerant enters the evaporator coil at a low pressure and

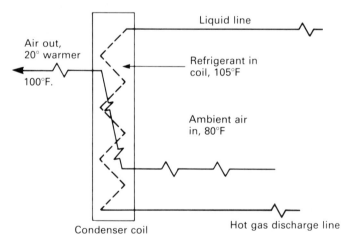

FIGURE 5–6
Relative temperatures of refrigerant in condensing coil and of air surrounding coil

as a mixture of liquid and vapor. Figure 5–7 shows the relationships of the refrigerant pressures and temperatures throughout the system.

5.6

Summary

The majority of air conditioning and heat pump systems are mechanical refrigeration systems. That is, they move heat by the process of refrigeration. As a result, a troubleshooter or a serviceperson must understand the refrigeration system in order to understand an air conditioning or a heat pump system.

A refrigeration system is a mechanical system that moves heat from a place where it is not wanted to a place where it is not objectionable. The heat is carried by a chemical compound called a refrigerant. A refrigerant is a material that has the property of boiling and condensing, or changing state, at relatively low pressure and low temperature.

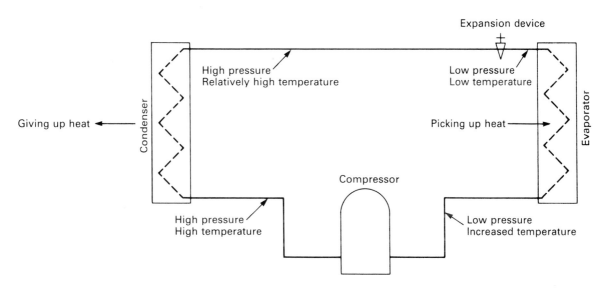

FIGURE 5–7

Pressure/temperature relationships of the refrigerant in a typical refrigeration system

As the refrigerant is circulated through the refrigeration system, the boiling and condensing temperature is controlled by the pressure. The pressure of the refrigerant in the system is controlled by the compressor and the expansion device. The refrigerant is maintained at a relatively low pressure in the evaporator, which is the part of the system where heat is picked up. It is maintained at a relatively high pressure in the condenser, which is the part of the system where heat is rejected.

There are two major sections in a refrigeration system: the low-pressure side and the high-pressure side. The major components in the low-pressure side of the system are the expansion device, the evaporator section, and the refrigerant lines. The major components in the high-pressure side are the compressor, the condensing section, and the refrigerant lines. Refrigerant lines connect the two sections of the system. Controls and other accessories control the flow of refrigerant and regulate the operation of the system.

5.7

Questions

1. Define *refrigeration system.*
2. Explain how a domestic refrigerator works.
3. What are the two main sections of a refrigeration system?
4. What is another name for the evaporator section of the system?
5. What is another name for the condensing section of the system?
6. Name the major components in the evaporator section.
7. Name the major components in the condensing section.
8. Name the refrigerant lines in a typical refrigeration system.
9. What is the most important characteristic of a refrigerant that enables it to be used for the transfer of heat in a system?

10. What is the main function of the compressor?
11. What is the main function of the expansion device?
12. Describe the term *boiling point temperature* as it relates to a refrigerant.
13. Define *saturated temperature*.

CHAPTER

Air Conditioning System

Air conditioning system

6.1

Introduction

An **air conditioning system** is a mechanical refrigeration system, as we saw in Chapter 5. The cool air is used for cooling a building or a part of a building. A refrigeration system is a major part of an air conditioning system, but an air conditioning system has some additional components and parts. The major additional parts are the evaporator blower, air distribution system, condenser blower, and air temperature controls. This chapter examines the total air conditioning system and its individual parts.

6.2

Total System

There are four major types of air conditioning systems. They are identified by the mediums that are cooled and by the mediums used to carry the heat away. The four types are air-cooled air conditioners, water-cooled air conditioners, air-cooled water chillers, and water-cooled water chillers. Many of the service procedures for these four types are similar. The air-cooled air conditioner is the most common type used for comfort air conditioning and is the only air conditioner covered completely in this text. A typical air-cooled air conditioning unit is shown in Figure 6–1.

FIGURE 6–1
Typical air-cooled air conditioning unit (Courtesy of Rheem
Air Conditioning Div.)

As described in Chapter 5, a refrigeration system has two main sections: the low-pressure side and the high-pressure side. The low-pressure side of the system has an expansion device, an evaporator coil, and some refrigerant lines. The high-pressure side of the system has a compressor, a condensing coil, and some refrigerant lines. An air conditioning system has the same main sections and components, as shown in Figure 6–2. In addition, each of the coils has a blower to move air across it. Also, an air conditioning system has a control system to regulate the cooling output and a distribution system for circulating the air into the spaces to be cooled.

Low-Pressure Side of a System

The **low-pressure side** of an air conditioning system starts at the expansion device and ends at the compressor inlet.

Low-pressure side

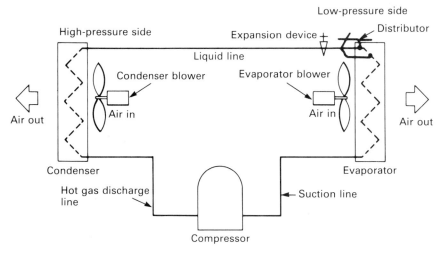

FIGURE 6–2   Schematic diagram of complete air conditioning system

It includes the expansion device, distributor, evaporator coil, evaporator blower, and suction line, as shown in Figure 6–2. A description of the parts, with the exception of the evaporator blower, follows. The blower is described in Chapter 18 along with the other parts of the air distribution system.

Expansion device

The **expansion device** provides one of the division points between the high- and low-pressure sides of the system. The device is located in the liquid line just before the line enters the evaporator coil. It creates a pressure drop, and consequent temperature drop, in the refrigerant. There are several different types of expansion devices used. The most common ones are described in Chapter 17.

Distributor

The refrigerant **distributor** is located between the expansion device and the evaporator coil. It looks like a nozzle, and it has one line entering and several smaller ones leaving it. The smaller lines feed refrigerant to the evaporator coil at several places. The distributor provides equal distribution of refrigerant throughout the coil. However, coils on smaller units may not have a distributor.

Evaporator coils

Evaporator coils are usually finned-tube coils. This type of coil is formed of copper tubes that have aluminum fins pressed onto them to increase the surface area exposed to the air. The coil is sized to carry the volume of refrigerant required for the cooling capacity of the unit and to accommodate the expansion of the refrigerant as it vaporizes.

The **suction line** is the refrigerant line that runs from the evaporator coil to the compressor. The suction line is larger than the other refrigerant lines since it must carry the refrigerant in the vapor state, that is, when the volume of the refrigerant is greatest.

Suction line

6.4

High-Pressure Side of a System

The **high-pressure side** of the system begins with the compressor and runs to the expansion device. It includes the compressor, hot gas discharge line, condenser coil, blower, and liquid line, as shown in Figure 6–2.

High-pressure side

The **compressor** is the second division point in the system where a pressure difference occurs between the high and low sides of the system. As the refrigerant goes through the compressor during an operational cycle, the pressure of the refrigerant gas is increased considerably. The temperature of the refrigerant also increases, because of the heat picked up from the motor windings and the heat of compression.

Compressor

Refrigerant leaves the compressor as a high-pressure, high-temperature gas through the **hot gas discharge line**. This line connects the compressor with the condenser coil.

Hot gas discharge line

The **condenser coil** is a finned-tube coil similar to the evaporator coil. It is sized to allow the quantity of refrigerant to flow as required for the cooling capacity of the unit.

Condenser coil

The **condenser blower** is an air mover that blows outside air across the condenser coil in order to remove heat from the refrigerant. Since the hot refrigerant vapor is at a higher temperature than the air, heat flows from the refrigerant to the air. The removal of heat causes the refrigerant gas to condense and become a liquid. The pressure is not changed appreciably, but the temperature is lowered in this process. The condenser blower is usually a propeller type of blower. The blower is sized

Condenser blower

to move the quantity of air required for the proper transfer of heat, depending on the temperature of the air and the refrigerant's condensing temperature.

6.5

Operation of a Typical Air Conditioning System

An air conditioning system, as we have seen, is made up of an expansion device, evaporator section, compressor, condensing section, and connecting refrigerant lines. These parts form a continuous loop through which a refrigerant circulates. The refrigerant picks up heat from the air that is blown across the evaporator coil and rejects that heat in the condensing coil. This section examines the sequence of events in the operation of an air conditioning system and describes the refrigerants used in typical systems.

Sequence of Operation

The sequence of operation for a typical air conditioning system begins when the temperature in the space where the thermostat is located drops below the set point of the thermostat. The cooling bulb in the thermostat closes an electric circuit in the control system, which sends an electric signal to the equipment. This electric signal actuates the compressor, the condenser blower motor, and the evaporator blower motor. With the condenser running, the refrigerant in the system circulates through the system. Air from the space to be cooled is blown across the evaporator coil by the evaporator blower. This air is then directed to the space to be cooled through the ducts in the distribution system.

As the air cools the space, the temperature in the space is reduced. When the temperature in the space drops below the thermostat set point, the thermostat's cooling bulb opens the electric circuit. With the electric signal interrupted, the air conditioning system shuts off.

The heat that is picked up from the cooled air is carried to the condensing system by the refrigerant. In the condensing section, the heat is rejected through the condensing coil as the condenser air is moved across the coil by the condenser blower.

Refrigerants

The **refrigerant** is a liquid that absorbs heat at a low pressure and temperature, and it rejects the heat at a high pressure and temperature. The refrigerant goes through a change of state from a liquid to a vapor at the low pressure/temperature condition, and from a vapor to a liquid at the high pressure/temperature condition.

    Several different refrigerants are used in air conditioning systems. Each one has different operating characteristics. The refrigerant selected for use in any given system is the one that works best for the particular application. The most common types of refrigerants used in comfort cooling applications are chemical compounds of halogen and carbon. A common name for these compounds is **Freon**. Freon is a generic name derived from the company that first manufactured them. Refrigerants are classified by number according to their chemical composition. Table 6–1 lists some common refrigerants and their number designations.

    The most important difference between the various refrigerants is the temperature at which they will change state at a specific pressure. A refrigerant must have a boiling point temperature, at a pressure that can be attained in the evaporator, that is lower than the temperature of the air to be cooled. It must also have a condensing temperature, at a pressure that can be attained in the condensing coil, higher than the temperature of the air used to pick up the heat.

Refrigerant

Freon

EXAMPLE

    A typical air-cooled air conditioning system may use any one of several refrigerants. In this example, we will assume that the system uses R–22. For given temperatures of the refrigerant in the evaporator and condenser coils, determine the refrigerant pressures that must be maintained in the system.

Refrigerant Number	Chemical Name	Chemical Formula	Molecular Weight	Boiling Pt. Degrees F.
11	Trichloromonofluoromethane	CCl_3F	137.4	74.8
12	Dichlorodifluoromethane	CCl_2F_2	120.9	−21.6
13	Monochlorotrifluoromethane	$CClF_3$	104.5	−114.6
13B1	Monobromotrifluoromethane	$CBrF_3$	148.9	−72.0
14	Carbontetrafluoride	CF_4	88.0	−198.4
22	Monochlorodifluoromethane	$CHClF_2$	86.5	−41.4
30	Methylene Chloride	CH_2Cl_2	84.9	105.2
40	Methyl Chloride	CH_3Cl	50.5	−10.8
50	Methane	CH_4	16.0	−259.
113	Trichlorotrifluoroethane	CCl_2FCClF_2	187.4	117.6
114	Dichlorotetrafluoroethane	$CClF_2CClF_2$	170.9	38.4
114A	Dichlorotetrafluoroethane	CCl_2FCF_3	170.9	38.5
142B	Monochlorodifluoroethane	CH_3CClF_2	100.5	12.2
152A	Difluoroethane	CH_3CHF_2	66.	−12.4
170	Ethane	CH_3CH_3	30.	−127.5
290	Propane	$CH_3CH_2CH_3$	44.	−44.2
500	Refrigerants 12/152A	CCl_2F_2/CH_3CHF_2	99.29	−28.0
717	Ammonia	NH_3	17.	−28.0
718	Water	H_2O	18.	212.
729	Air	——	29.	−318.
744	Carbon Dioxide	CO_2	44.	−109. (Subl.)
764	Sulphur Dioxide	SO_2	64.	14.0
1150	Ethylene	$CH_2 = CH_2$	28.0	−155.0
1270	Propylene	$CH_3CH = CH_2$	42.1	−53.7

Courtesy of Carrier A.C. Group

TABLE 6–1   Refrigerants

Solution

By determining the temperatures we want for the refrigerant in the evaporator and condenser coils, we can determine the refrigerant pressures we will have to maintain in the two parts of our system. In a typical air conditioning system, the temperature of the air leaving the evaporator coil should be about 55 to 60 degrees Fahrenheit. There will have to be about a 20-degree difference between the air and the refrigerant to allow for heat transfer. So, the refrigerant temperature should be about 35 to 40 degrees Fahrenheit. Checking a pressure/temperature chart similar to the one in Figure 5–5, we find that we need a pressure of about 61 to 68 pounds per square inch for this temperature range. This pressure, then, is the pressure we should have in the low-pressure side of the system.

Outside air is used for the condensing medium in an air-cooled air conditioning system. The outside temperature will vary, but it can easily go as high as 110 degrees Fahrenheit in some localities. Again, the temperature difference between the refrigerant and the air should be about 20 degrees to allow for good heat transfer. So, the refrigerant in the condensing coil should be at least 130 degrees Fahrenheit. Checking the pressure/temperature chart again, we find that we need a pressure of 181 pounds per square inch to achieve this temperature. This pressure is the pressure we must maintain in the high-pressure side of the system.

6.6

Controls

Refrigerant flow within an air conditioning system is controlled by the expansion device, by the compressor, and, to a lesser extent, by the temperature of the condenser and evaporator mediums. The **control** of the cooling output of the unit, over a given period of time, is achieved by cycling the compressor on and off. This cycling of the compressor is done in one of two ways. In the first method, cycling is achieved by controlling the operation of the compressor through a thermostat. In the second method, cycling is achieved by controlling the refrigerant pressure through a solenoid valve on the liquid line and then using the low-pressure control to cycle the compressor on and off. In either case, the compressor is cycled on and off in order to provide a variable cooling output over a fixed period of time.

Figure 6–3 shows a typical control system using thermostatic control. This typical control system has two types of controls: operating controls and safety controls. The **operating controls** include the control system trans-

Control

Operating controls

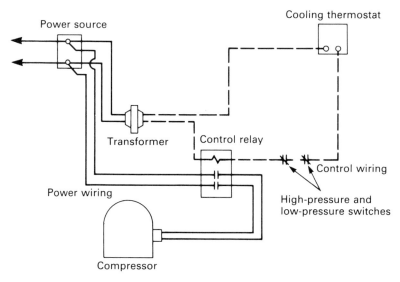

FIGURE 6-3
Schematic diagram of control system for typical air conditioning system

Safety controls

Thermostat

Electric relay

former, the thermostat, and the relay (or contactor) that controls the compressor. The **safety controls** include pressure switches, motor overloads, and other controls that protect the parts of the system from damage in case of a mechanical malfunction.

Details of the construction and function of the various controls is given in Chapter 10. A brief description of the sequence of operation of a typical unit follows.

When the **thermostat**, which is a temperature-actuated electric switch, calls for cooling, an electric relay is energized. The **electric relay** is an electric switch that allows one circuit to control another. The relay either controls the compressor directly or, in larger units, controls a contactor or magnetic starter that, in turn, controls the compressor. The thermostat simultaneously energizes other relays that turn on the condenser blower and the evaporator blower. When the air temperature in the area being cooled is the same as the set point temperature of the thermostat, the thermostat is satisfied, and the electric control circuit is opened. The relays that control the compressor and the blower motors are then de-energized, so the unit is shut off.

A high-pressure switch monitors the pressure in the high-pressure side of the system during operation. A low-pressure switch monitors the pressure in the low side. If the pressure in the high side exceeds a set point pressure—that is, a pressure that is a safe operating pressure for the unit—then a set of electric contacts in the control circuit opens, and the unit is shut off. If the pressure in the low-pressure side of the system drops below a predetermined setting considered safe for operation of the unit, then the same thing happens.

The control circuit for the unit is wired through the contacts in each of these pressure switches. Other safety controls, such as motor overloads, are an integral part of the motors. **Motor overloads** are safety switches that turn the motor off in case of overheating or overloading of the motor.

Motor overload

6.7

Distribution System

While all of the mechanical sections of the air conditioning system are necessary to its operation, the cool air produced by the system has to be distributed to a point of use before the function of the system is complete. For distribution of the cool air, a blower, supply and return ducts, supply and return registers, and grilles are used. This part of the total system is called the **distribution system**. Each part is discussed in the following subsections.

Distribution system

Evaporator Blower

The blower that moves air over the evaporator coil and through the distribution system is called the **evaporator blower**. It is usually a centrifugal blower. A **centrifugal blower** has a rotating wheel that draws air in at the center and discharges it out the side. Figure 6–4 shows a centrifugal evaporator blower.

Evaporator blower
Centrifugal blower

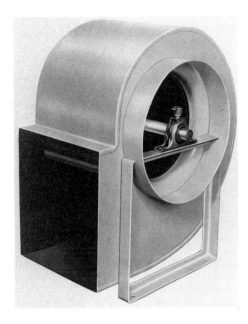

FIGURE 6–4

Centrifugal blower used as evaporator or indoor blower (Courtesy of Barry Blower, a Marley Company)

The air-moving capabilities of a centrifugal blower are such that a given size of blower will move a fixed volume of air against the friction in the ducts most efficiently. The air is either blown through or drawn through the evaporator coil by the blower. The air then goes through the supply ducts to the spaces to be cooled. Finally, the air is returned from the cooled spaces through the return ducts to be processed again.

Ducting

Ducts are used in an air distribution system to carry the conditioned air to the spaces where cooling is desired. The ducts are sized to carry the volume of air required at a minimum friction loss. The air is moved through the duct system by the blower.

A typical duct system is shown in Figure 6–5. The duct system includes both supply ducts and return air ducts. The duct system is made up of a main duct, called a trunk, and branch ducts that run from the trunk to the registers or grilles.

Registers, Grilles, and Dampers

In an air distribution system, the air to cool the various areas in the building is distributed to those areas by **supply registers**. A typical supply register is shown in Figure 6–6A. These registers are located on the floors,

Ducts

Supply registers

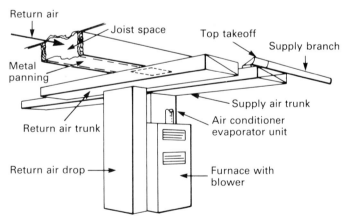

FIGURE 6–5

Typical duct system

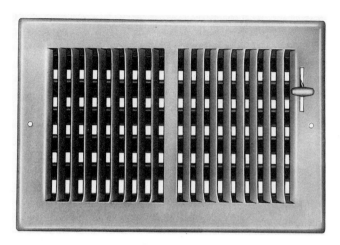

A. Supply Register

B. Return Air Grille

FIGURE 6–6   Supply register and return air grille for air distribution system (Courtesy of Lima Register Co.)

walls, or ceilings of the area. They are connected to the supply ducts either directly or by branch ducts. They are sized to deliver the quantity of air required in each space for the cooling load. The air is delivered through the registers at a velocity and in a pattern that will cool the room best. The volume of air is controlled by **dampers** in the registers themselves or in the ducts or branches.

Dampers

In a distribution system, air from the rooms is returned to the blower by the **return air system**. The return air system starts at the return air grilles in the cooled rooms. A **return air grille**, shown in Figure 6–6B, is a decorative covered opening that leads through the branches to the return air duct system. The return air duct is connected to the return air side of the blower.

Return air system

Return air grille

Summary

An air conditioning system is a mechanical system that moves heat from one place to another, generally from a place where it is not wanted to a place where it can be rejected. The heat is carried by a refrigerant that is circulated through a piping system. The temperature of the refrigerant in the various parts of the system is controlled by the pressure. The pressure of the refrigerant is, in turn, controlled by the components of the system. There is a high-pressure side of the system and a low-pressure side. The temperature of the refrigerant in the high side of the system is relatively warm, and the temperature in the low side is relatively cool.

An air-cooled air conditioning system is a system designed to cool air. Air is blown across a coil located in the low-pressure side of the system, and the air is cooled as heat is removed from it by the refrigerant in the coil. The refrigerant then goes through a compressor, and its pressure and temperature are raised. The refrigerant next goes to a coil in the high-pressure side of the system. Outside air passes over this coil, and heat is rejected by the refrigerant to the outside air, because the outside air is cooler than the refrigerant in the coil. As the refrigerant is cooled, it condenses to a liquid. This liquid refrigerant then flows from the coil to the expansion device, which is located just ahead of the low-side coil in the system. The expansion device provides a pressure drop in the refrigerant, and the cycle begins again.

After the air is cooled in the air conditioning part of the system, the distribution system circulates the air to the areas or rooms that it will cool. The distribution system is composed of a blower, supply and return air ducts and branches, supply registers, and return air grilles. Balancing dampers located in the supply registers or in the branch ducts control the flow of air into each room or space to be cooled.

Questions

1. Describe an air conditioning system, and explain the difference between an air conditioning system and a refrigeration system.
2. Name the four types of air conditioning systems.
3. Name the components in the low-pressure side of an air conditioning system.

4. Why are finned-tube coils used as evaporator and condenser coils in an air conditioning system?
5. Name the components in the high-pressure side of an air conditioning system.
6. What is the main difference between the various refrigerants used in air conditioning systems?
7. What is the saturated temperature of refrigerant R–22 if the pressure of the refrigerant is 84.0 pounds per square inch?
8. Name and describe the two methods of control used for most air conditioning systems.
9. A typical control system has two types of controls. Name them.
10. Outline the sequence of operation of an air conditioning system.
11. Describe the function of a high-pressure switch.
12. Describe the function of a low-pressure switch.
13. Name the main parts of a distribution system.

CHAPTER

Troubleshooting Charts

7.1

Introduction

Troubleshooting is the process of checking an air conditioning or a heat pump system that is not functioning properly in order to determine the cause of the problems in the system. As mentioned earlier, even though the physical acts of troubleshooting and service are often performed simultaneously, they are considered separately in this text.

A troubleshooting call on an air conditioning or a heat pump system is usually initiated by a complaint that the system is not working the way it should. In the troubleshooting procedure, a technician analyzes the complaint to determine the symptoms of trouble. Then, he or she analyzes the symptoms to identify possible causes. A method for systematically checking the sections, components, and parts of a system—as they are related to the complaints, symptoms, and causes—simplifies the identification of faulty parts. Such a systematic approach is presented in this chapter.

Troubleshooting chart

A **troubleshooting chart** is a listing of the procedures to be followed in identifying the part of the system involved when a problem occurs. The chart also helps a technician to identify the actual parts involved. A troubleshooting chart helps a technician to find problems quickly and efficiently. Thus, use of the chart can save time and money for both the troubleshooter and the owner of the equipment. This chapter presents portions of a troubleshooting chart and describes how the chart is used.

7.2

Systematic Approach

To troubleshoot an air conditioning or a heat pump system most efficiently, a troubleshooter should follow a step-by-step process of checking the system for problems. This step-by-step process is a **systematic approach**.

Three steps are involved in a systematic troubleshooting process. The process begins when a customer complains that a system is not working the way it should. The first step is to analyze the complaint. The second step is to identify the symptoms related to the complaint. The third step is to analyze the symptoms in order to identify the cause of the failure. In other words, the following three steps are used in the systematic approach to troubleshooting an air conditioning or a heat pump system.

1. *Analyze the complaint.* Most complaints are stated in general terms and do not include specific data about the operation of the system. The troubleshooter must determine what the complaint means in terms of actual system operation.
2. *Identify the symptoms related to the complaint.* Analysis of a complaint allows a troubleshooter to describe the performance of the unit in terms of normal performance.
3. *Analyze the symptoms to identify causes.* Analysis of the symptoms identifies the cause of the problem and identifies the parts of the system that are not operating properly.

Systematic approach

Complaints

Each troubleshooting call originates with a complaint. The **complaint** is related to how good a job of cooling or heating the system is doing. The complaint usually originates with someone working in the building in which the system is installed. Often, the person registering the complaint is not familiar with the actual operation of the air conditioning system. Thus, the complaint is usually

Complaint

stated in general terms and does not include specific information about the system.

If the system involved is an air conditioning system, then the complaint is usually one of the following: no cooling, not enough cooling, too cool, noisy system, or too expensive to operate. If the system is a heat pump system, then the complaints about the heat may be as follows: no heat, not enough heat, or too hot. The complaints are generally related to the total system and not to one particular part of it.

The following list summarizes and describes the eight basic complaints for air conditioning and heat pump systems.

1. *No cooling:* If a system operator is unable to maintain as low a temperature as is desired with the system, the complaint will usually be that the system is not working at all.

2. *Not enough cooling:* In some cases, the operator may feel that the amount of cooling produced by the system is insufficient and that more cooling effect should be felt.

3. *Too cool:* In some cases, the complaint may be that the system is keeping a space cooler than it should be.

4. *No heat:* If both the heat pump and the auxiliary heat system become inoperative, the system will not produce any heat. This problem is usually caused by a general power failure or by a failure in the electric power supply system within the system.

5. *Not enough heat:* This problem is usually caused by failure of the auxiliary heat system. In this case, the heat pump is not capable of supplying enough heat to satisfactorily heat the entire building.

6. *Too hot:* In some cases, the complaint may be that the heat pump system is keeping a space hotter than it should be.

7. *Noisy system:* Mechanical problems within a system can generate noise. And in many types of building uses, noise from the air conditioning or heat pump system can be disturbing or disruptive to activities in the building. In these cases, a complaint of a noisy system may be registered.

8. *Too expensive to operate:* Some types of mechanical failure can result in high energy costs. In some cases, general discontent with the system may result in a complaint of high operating costs.

When a complaint is received, a service technician should find out as much as possible about how the system is operating from the person making the complaint. Most complaints are made by people who are not acquainted with the technical operation of the air conditioning system. Often, however, they can give much helpful information about the system. For instance, if the original complaint is that the unit doesn't cool the building, subsequent questioning may reveal that the air from the registers doesn't feel as cool as it once did. This kind of information helps the troubleshooter identify the problem more specifically than the original complaint does.

After getting as much information as possible from the person making the original complaint, the troubleshooter should ask for additional information from others who may be familiar with the system. A plant manager or an operating engineer may often notice such things as the frequency of the operating cycles, and he or she may be able to give some indications of the specific performance of the system. A technically trained person will usually be able to give specific information, not just general data.

After as much data as possible have been gathered relative to the original complaint, the next step is to look for symptoms in the operation of the system.

Symptoms

After all the data concerning a complaint are received, the troubleshooter must analyze the complaint and identify the symptoms related to the complaint. This analysis is done by investigating the system and observing how it is operating. A **symptom** is a sign or indicaton of something that is wrong. For example, a symptom related to the complaint of "not enough cooling" is no air coming from the registers.

Symptom

While complaints are related to the total system, symptoms are related to one of the major components in the system. Some of the most common symptoms found in an ailing system are listed in Table 7–1.

As the troubleshooter relates complaints to symptoms, he or she should also start thinking in terms of the major components of the system. The major components related to specific symptoms are listed in Table 7–2.

Here are some examples of how complaints are related to symptoms.

- If a complaint of "no cooling" is received, one symptom may be no air coming from the supply registers. This symptom may indicate that something is wrong with the blower in the air distribution part of the system.
- If a complaint of "not enough cooling" is received, one symptom may be that the air from the registers is warm instead of cool. This symptom may indicate that something is wrong with the refrigeration part of the system.
- If a complaint of "too cool" is received, one symptom may indicate problems in the design of the air distribution system.
- If a complaint of a "noisy system" is received, one symptom may indicate mechanical problems somewhere in the system.
- If a complaint of "too expensive to operate" is received, one symptom may be that the control system is keeping the cooling unit running too long on either a cooling or a heating cycle.

After investigation, the most probable symptom related to the complaint is determined.

Causes

Cause

With a specific symptom of a system malfunction identified, the next step is to determine the possible **cause** of the malfunction, that is, the part of the system not operating properly. Each set of symptoms related to a

Complaint	Symptom
No cooling	No air coming from the supply registers
	Not enough air coming from the supply registers
	Warm air coming from the supply registers
Not enough cooling	Room air temperature higher than set point temperature of the thermostat
	Warm air coming from the supply registers
Too cool	Room air temperature lower than the thermostat setting
	Cool drafts in the area
No heat	No air coming from the supply registers on a call for heat by a heat pump
	Cool air coming from the supply registers on a call for heat by a heat pump
Not enough heat	Room air temperature lower than heating set point of the thermostat
	Cool air coming from the supply registers on a call for heat
Too hot	No air coming from the registers on a call for cooling
	Not enough air coming from the registers on a call for cooling
	Warm air coming from the registers on a call for cooling
Noisy system	Popping or ticking noises from registers
	Scraping or squeaking noises from registers
	Noisy compressor
	Noisy condenser unit
	Noise from refrigerant lines where they go through parts of building
Too expensive to operate	Unit runs constantly
	Unit needs frequent service
	Unit will not come out of defrost

TABLE 7–1   Common Symptoms for Complaints Received about Air Conditioning or Heat Pump Systems

Symptom	Major component
No air coming from the supply registers	Evaporator blower Balancing dampers Fire dampers
Warm air coming from the supply registers	Compressor Expansion device Condensing unit
Room air temperature higher than set point temperature of the thermostat setting	Compressor Expansion device Condensing unit
Air too warm from the supply registers	Compressor Expansion device Condensing unit Air distribution system
Room air temperature lower than the thermostat setting	Equipment undersized Air distribution system
Cool drafts in the area	Air distribution system
Popping or ticking noises from registers	Air distribution system
Scraping or squeaking noises from registers	Air distribution system
Noisy outdoor unit	Compressor Condenser blower
Vibration or rattling in building	Refrigerant lines
Unit runs constantly	Controls
Unit needs frequent service	General equipment

TABLE 7–2   Major Components of System Related to Symptoms

system malfunction has a certain number of possible causes related to it. Table 7–3 lists some of the causes that can be related to the symptoms for a typical system. The troubleshooting charts covered later in this chapter list other causes a troubleshooter can look for.

As indicated in Table 7–3, if the symptom is a noisy condenser blower, the cause could be the condenser fan blower blades striking the fan housing. The cause could also be worn-out motor bearings or blower bearings. Investigation will locate the actual cause.

Eventual identification of the faulty part causing a system malfunction is a result of, first, analyzing the complaint; second, defining the symptoms that are re-

Symptom	Cause
No air coming from the supply registers	No electric power to the air conditioning equipment
	Evaporator blower not operating
	Balancing dampers closed
	Fire dampers closed
Not enough air coming from the supply registers	Balancing dampers partially closed
	Evaporator blower cycling on overload
	Blower belt slipping
	Dirty filters
	Dirty evaporator coil
Warm air coming from the supply registers	Compressor not operating
	Condenser fan motor not operating
	Condenser coil obstructed
Room air temperature higher than set point temperature of the thermostat	Faulty thermostat
	Anticipation incorrectly set
	Cooling equipment undersized
	Supply registers too small
	Return air grilles too small
	Ductwork undersized
	Dirty filters
	Dirty evaporator coil
Air too warm from the supply registers	Compressor cycling
	Condenser blower not working properly
	Condenser coil dirty
	Evaporator blower belt slipping
	Dirty filters
	Dirty evaporator coil
	Supply ductwork not insulated in warm area
	Return air registers in cold area
Room temperature lower than the thermostat setting	Thermostat out of calibration
	Thermostat contacts welded
	Thermostat circuit wiring shorted

TABLE 7–3   Causes of Problems Related to Symptoms

Symptom	Cause
Cool drafts in the area	Control relay contacts welded
	Cooling equipment oversized for job
	Supply registers poorly located
	Wrong type of supply registers used
	System never balanced
Popping or ticking noise from registers	Metal ductwork not braced adequately
	Ductwork too short or not insulated
Scraping or squeaking noises from registers	Blower wheel rubbing on blower housing
	Blower bearing worn out
	Evaporator blower motor bearing worn out
	Evaporator blower belt slipping
Noisy compressor	Compressor bearing worn out
	Broken valves in compressor
Noisy condensing unit	Condenser blower blades hitting housing
	Condenser blower bearings worn out
	Condenser blower motor bearings worn out
Noise from refrigerant lines where they go through parts of building	Lines not insulated from structure
	No vibration isolator on hot gas discharge line
Unit runs constantly	Thermostat out of calibration
	Thermostat contacts welded
	Control circuit wiring short-circuited
	Control relay contacts welded
	Cooling equipment undersized for job
	Dirty filters
	Dirty evaporator coil

TABLE 7–3   *(continued)*

Symptom	Cause
Unit needs frequent service	System improperly designed System installed improperly System not cleaned up properly after a burnout Inadequate maintenance

TABLE 7–3   (continued)

lated to that complaint; and third, identifying the cause of the complaint. Symptoms are usually found to be related to major components, and faults are found to be related to parts. A troubleshooter looks for symptoms and faults, while a serviceperson looks for components and parts. Table 7–4 is a portion of a service chart that identifies causes and the parts that are at fault.

7.3

Using a Troubleshooting Chart

A troubleshooting chart is a graph or chart that lists complaints, symptoms, and causes of problems. The causes are cross-referenced on the chart with the parts of the system. The cross-referencing is arranged so that the user can identify the parts of the system that are

Cause	Part
Faulty thermostat	Thermostat
Anticipation	Thermostat
Equipment undersized	Total system
Supply registers too small	Registers
Return air grilles too small	Ducts and grilles
Ductwork undersized	Ducts
Dirty filters	Filters
Dirty evaporator coil	Coil

TABLE 7–4   Portion of Service Chart Showing Parts of System Related to Causes

causing problems. By using a troubleshooting chart, a technician can quickly identify faulty sections and parts. A portion of a troubleshooting chart is presented in Table 7–5. Complete troubleshooting charts for both air conditioning and heat pump systems are given in the Appendix. Use of these charts will help a troubleshooter follow the systematic troubleshooting steps discussed earlier. By following the charts, a troubleshooter is carried naturally through a checkout procedure for a malfunctioning system.

In troubleshooting charts, the complaints are listed in the left column of each section. These complaints are the major complaints received concerning system operation. Each chart also has three other columns. The second column lists the symptoms related to the complaints. The third column lists the causes related to the symptoms. The fourth column, on the right of the chart,

Complaint	Symptom	Cause	Section, component, or part
Noisy system	Popping or ticking noise from registers	Metal ductwork not braced properly	Air distribution system
	Scraping or squeaking noises from registers	Evaporator blower wheel rubbing on housing	Evaporator section
		Blower bearings worn out	
		Blower motor bearings worn out	
		Blower belt slipping	
	Noisy compressor	Refrigerant flooding back to compressor	Expansion device Evaporator blower Refrigerant charge
		Compressor bearings worn out	Compressor
		Broken valves in compressor	
		Low oil level in compressor	

TABLE 7–5   Portion of a Troubleshooting Chart

lists the major section, component, or part related to the causes.

To use a troubleshooting chart, locate the row in the first column that lists the complaint received about the system. In the second column, find the row listing the most obvious symptom for that complaint, according to your investigation of the system. In the third column, identify the probable cause of the symptom. Finally, in the fourth column, identify the major section, component, or part related to the cause.

EXAMPLE

Use the air conditioning troubleshooting chart in the Appendix to investigate a complaint of "no cooling."

Solution

We locate the "no cooling" row in the first column of the chart. In the second column, we find the symptom that seems to fit the situation. A number of possible symptoms are listed. We look for the one that seems to fit the particular complaint we received. For instance, if no air is coming from the registers, we find this symptom in the chart. In the third column, across from the symptom of "no air coming from supply registers," several possible causes are shown. Further investigation of the system will help us determine which of these causes is the most probable cause of the problem in the system. For instance, a visual inspection of the system may show us that the evaporator blower is not operating. Further investigation will show us why the blower is not running.

The use of troubleshooting charts assists a technician in identifying faulty parts of a system. Service charts, which are used by servicepeople, are an extension of troubleshooting charts. The systematic process of troubleshooting a system is made quicker and easier by the use of these charts.

7.4

Summary

A systematic method of troubleshooting an air conditioning or a heat pump system has three steps. The first step is to analyze a complaint concerning system operation. The second step is to identify the symptoms related to the complaint. The third step is to analyze the symptoms in order to identify the cause of the problem.

Complaints of system operation are usually general in nature and are related to the effect of a malfunction rather than to the malfunction itself. A troubleshooter must analyze the complaint and think of it in terms of symptoms that are indicated. These symptoms, in turn, are used to identify the causes of the problems within the system.

A troubleshooting chart lists the most common complaints received, the symptoms that may be related to those complaints, and the most probable causes related to those symptoms. Use of a troubleshooting chart helps a service technician follow a systematic process when troubleshooting air conditioning or heat pump systems.

If a troubleshooter follows a systematic method of searching for problems when investigating a complaint concerning the operation of an air conditioning or a heat pump system, much time and effort can be saved. A haphazard or general search for faulty parts takes a much longer time and will be more costly for both the troubleshooter and the owner of the system.

7.5

Questions

1. Define a *troubleshooting chart*.
2. Describe a step-by-step process for troubleshooting.
3. What are the three steps in a systematic process of troubleshooting?
4. Describe a complaint as related to troubleshooting.

5. When a complaint is received concerning the operation of a system, what is the first thing a troubleshooter should do?
6. Describe a symptom as related to troubleshooting.
7. How is a cause related to a symptom?
8. Describe a typical troubleshooting chart.

Service Charts

8.1

Introduction

In this chapter a systematic procedure to be followed in servicing an air conditioning or a heat pump system is discussed. Parallels between troubleshooting and servicing will be recognized as these procedures are discussed. The serviceperson's work starts where the troubleshooter's ends.

Troubleshooting is the process of systematically identifying general problems within a faulty air conditioning or heat pump system. Service is the process of identifying the specific components and parts related to the problems and then repairing or replacing the faulty parts to correct the problem. Troubleshooting and servicing are so alike that it is difficult to think of them

Troubleshooting chart	Service chart
Complaint: General statement concerning the operation of the system	Major sections: High-pressure side of the system and low-pressure side of the system
Symptom: Results of system operation	Components: Major parts of each of the major sections
Causes: Reasons for improper system operation	Parts: Individual parts of the various components of the system

TABLE 8–1   Elements in Troubleshooting and Service Charts

separately. For the sake of clarity, even though many of the steps in the two processes are carried out simultaneously, they should be thought of as two separate processes.

Service charts are also described in this chapter. A service chart is similar to the troubleshooting charts discussed in Chapter 7. A **service chart** outlines the procedures to be followed in servicing a typical air conditioning or heat pump system. The use of the service chart is also explained in this chapter.

Service chart

Systematic Approach to Service

In the troubleshooting process, complaints related to the operation of the entire system are reduced to symptoms, and symptoms are reduced to causes. In the service process, symptoms are related to components, and causes are related to the parts of the system. The components of the system are investigated in order to find the parts that are at fault. Table 8–1 shows the relationships between troubleshooting and service charts.

Identification of the faulty parts is the first part of the service process, and repair or replacement of the parts is the second part. It is as important to follow a systematic procedure in servicing an air conditioning or a heat pump system as it is in troubleshooting a system, and for the same reasons. Time, effort, and money are saved if a service technician follows a step-by-step procedure when identifying and repairing faulty parts.

Major Sections of an Air Conditioning System as Related to Service

As discussed in Chapter 6, an air conditioning (or heat pump) system is made up of two **major sections**: the low-pressure side and the high-pressure side. In turn, each section contains **components**. The low-pressure side of the system contains the expansion device, the evap-

Major sections

Components

Parts

orator section, refrigerant lines, accessories, and controls. The high-pressure side of the system contains the compressor, the condenser section, refrigerant lines, accessories, and controls. Figure 8–1 illustrates these parts of a typical system.

Each of the components of the system is, in turn, made up of individual **parts**. For instance, the evaporator section has a distributor, a coil, a blower, and various accessories and controls. The condensing section has a compressor, a blower, a coil, and various accessories and controls. In addition, each of these parts has many individual parts. For example, the compressor has a motor, valves, rotors or pistons, bearings, and other parts that affect its operation.

Some parts may be duplicated in different components in the system. For instance, electric motors are found in each of the major sections of a system. There is a compressor motor, an evaporator blower motor, and a condenser blower motor. Some electric power and unit controls are duplicated in different parts of the system. Both a main disconnect and a unit disconnect are used on almost all systems.

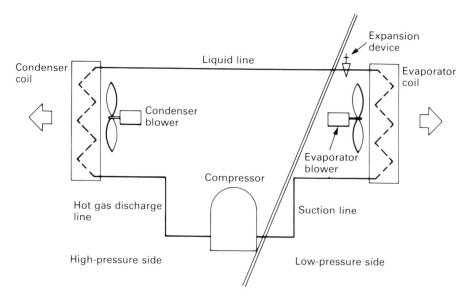

FIGURE 8–1

Main components of air conditioning system in relation to major sections

In addition to components and parts, air conditioning or heat pump systems also have a distribution system. And this part of the system includes many more parts.

To facilitate the servicing of a system, we can separate the components and parts into **divisions according to function**, as follows:

Divisions according to function

- Electric power supply components
- Controls
- Electric motors
- High-pressure side of the system
- Low-pressure side of the system
- Distribution system

These divisions by function will be used throughout the rest of this text in discussing service procedures. These major divisions are identified in Figure 8–2. The components shown in Figure 8–2 were described in Chapter 6.

The first step to take in the servicing procedure is the identification of the individual part that is causing trouble. Here, again, a systematic method of checking should be used. A part that has been identified as the

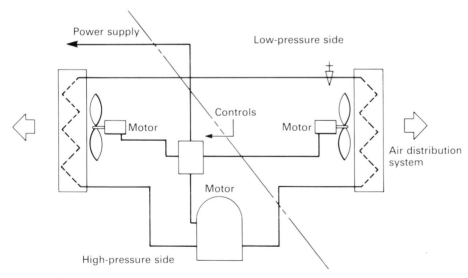

FIGURE 8–2
Major divisions of air conditioning systems by function

possible cause of trouble should be checked out independently of the other parts of the component. The second step in the servicing procedure is to repair or replace faulty parts. The proper procedure for checking the most common parts in a typical air conditioning or heat pump system is described in following chapters.

8.4

Using Service Charts

A service chart indicates the order in which a serviceperson should check the major sections, components, and parts of an air conditioning or a heat pump system. The chart helps the technician to locate the problem when a unit is not operating as it should. A service chart outlines a systematic procedure for a serviceperson to follow.

Service charts come in many different forms, but all of them have the same objective. That objective is to lead the serviceperson to the source of the trouble in the fewest possible steps. After some experience, a serviceperson may not have to use a chart for every service call. But a good service technician will always follow the procedure outlined by a service chart.

In a common form of service chart, the starting point is a complaint about the operation of the system. The chart leads from the complaint to the possible problem. Then, it leads from the problem to the major section of the system in which the problem originates. Next, it directs the serviceperson to the components in which the problem exists. Finally, it leads from the component to the faulty part (or parts) within the component. Some charts include a brief explanation of how the technician can check the part for failure. Other charts give a reference showing where the information for checking the parts can be found.

A set of service charts is given in the Appendix. A sample of a service chart is presented in Table 8–2. In the service charts in this text, the complaint is listed in the left column. A list of possible problems is presented in the second column. Major sections of the system are listed in the third column. These sections are the same

Complaint	Problem	Major section	Component	Part	Reference
No cooling	No air coming from supply registers	Electric power to unit	Main power supply	Branch; disconnect	Chap. 9
				Wiring	Chap. 9
				Fusing or breakers	Chap. 9
				Unit; disconnect	Chap. 9
				Wiring	Chap. 9
				Fusing or breakers	Chap. 9

TABLE 8–2   Portion of a Service Chart

sections listed earlier in this chapter. Components are listed in the fourth column, across from the major sections they are related to. The parts are listed in the fifth column. Finally, in the sixth column on the right, a chapter reference is given. These references identify the chapter in which a description is given for checking that particular part. Conscientious use of a service chart helps a serviceperson identify faulty parts in a system in a more efficient manner than any other procedure.

To use the service charts in the Appendix, start with the complaint received. Reference to the chart shows a number of problems that could be related to each complaint. Identify the problem in the chart. Next, find the major section of the system related to the problem. Now, check the components within the selected section. Upon investigation, note the component that appears most likely to be at fault. Finally, investigate the parts within the components and identify the faulty ones. An outline of the procedure just described is illustrated in Figure 8–3.

EXAMPLE

A complaint of "no cooling" is received concerning an air conditioning system. Find the part causing the problem.

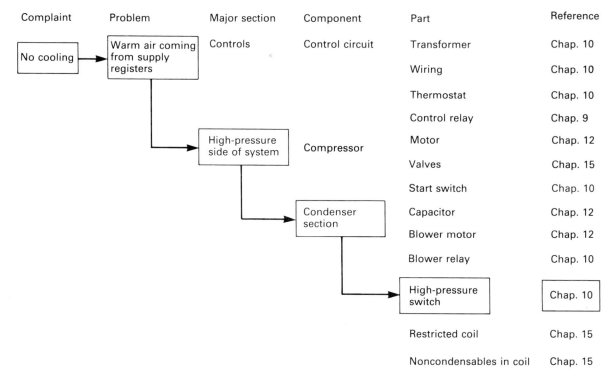

FIGURE 8–3
Flowchart showing use of service chart

Solution

In the service chart for air conditioning given in the Appendix, we choose the row with the heading "no cooling." The chart shows that there are two possible problems for this complaint. The first problem is that there is no air coming from the supply registers. The second problem is that warm air is coming from the supply registers. Examination of the system quickly indicates which of the two conditions exists.

Suppose the first problem is found. If there is no air coming from the registers, then we look in the "major section" column. This column shows that the first section of the system to investigate is the electric power section. Going across the chart to the "component" column shows that the main power supply to the unit is the first component to check.

For a check of this component, we continue across the chart to the "part" column. This column lists the parts that should be checked for failure. The parts are listed in the order in which they should be checked. That is, the first part in the list is the one most apt to be at fault, and the other parts are listed in the order in which they are apt to be at fault. The parts can be checked as described in the referenced chapter of this text. Figure 8–4 shows the procedures just discussed marked on a page of the service chart.

Specific Part Check

The last step in the service process, before a part is replaced, is to check the part that is suspected of being at fault. All of the steps that have been covered in the troubleshooting and servicing procedures to this point are only preliminaries to this final step.

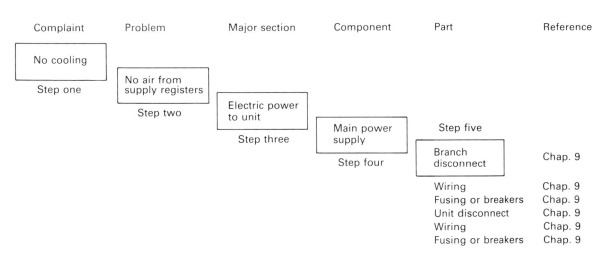

FIGURE 8–4
Example service procedures marked on chart

Each part is checked in a step-by-step process. The actual procedure varies somewhat depending on the type of part being checked. Thus, the procedure for a mechanical device is different from the procedure for an electric device. But the checking out process is similar. Tools, instruments, and the physical process used vary according to the part. The following chapters in this text describe the parts and explain how a technician should check each one for proper operation.

8.6

Summary

Equipment service is an extension of troubleshooting. The information found in troubleshooting a faulty system is used to identify the faulty parts in the system. After the faulty parts are identified, a serviceperson repairs or replaces them. Service begins where troubleshooting leaves off.

The major sections of an air conditioning or a heat pump system are the low-pressure and high-pressure sides of the system. The expansion device, the distributor, the evaporator section, the suction line, and various accessories and controls are all part of the low-pressure side of the system. The compressor, the condensing section, the hot gas discharge line, the liquid line, and various accessories and controls are all part of the high side. Individual controls are found within each of these parts of the system.

To service a system most efficiently, a technician should follow a systematic procedure. A service chart provides the best outline of the procedures to be followed in servicing a system. A service chart provides a serviceperson with an outline that leads from the complaint received, to the major sections of the system, to the components within the section, and finally to the part (or parts) that has failed. A good service chart also directs the serviceperson to information explaining how each part can be checked. Conscientious use of a good service chart enables a serviceperson to find a faulty part in a system quickly and easily.

8.7

Questions

1. Explain why a service chart is used.
2. In the service of a system, the symptoms described in troubleshooting can be related to _____, and causes can be related to _____.
3. Name the six functional categories of the components and the parts of a system.
4. Describe a service chart.
5. Why is the method of checking a system as outlined by a service chart a good method to use?
6. A service chart helps to identify individual parts that may be faulty. But what must then be done to finish servicing a system?

Electric Power Supply System

9.1

Introduction

Most modern air conditioning or heat pump systems are operated and controlled by electric power. The electricity to furnish this power is carried by wires in separate circuits to each device in the system. The power is controlled by electric switches, relays, and other control devices. The wires, switches, relays, and other electric controls are the subject of this chapter.

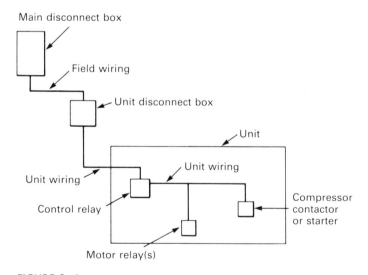

FIGURE 9–1
Electric power supply system for air conditioning or heat pump system

The electricity used to operate the equipment in an air conditioning or a heat pump system comes into the building from the electric utility power lines in the area. Inside the building, the power lines enter a main control panel. This panel has a disconnect switch and fuses for the incoming power line. Branch circuits run from the main panel to separate loads in the building, such as the air conditioning or heat pump units. Each of these branches also has a disconnect switch and fuses or circuit breakers at the main panel. There are unit disconnects and fuse boxes on each unit, also. The electric power circuit enters the equipment from the unit disconnect box.

This chapter describes each of these parts of the electric power supply system and explains the methods used to service them.

9.2

Parts of the System

The **electric power supply** section in an air conditioning or a heat pump system consists of (1) the wiring that carries electric current, (2) devices that allow control of the current, and (3) controls that regulate the flow of the current to the various parts of the system. This part of the system, then, includes the wiring, disconnects, fuses or overcurrent protection devices, and relays, starters, or contactors.

Electric power supply

A list of the **eight common controls** in the system follows.

Eight common controls

1. Main disconnect
2. Main fuses or circuit breakers
3. Wiring
4. Unit disconnects
5. Unit fuses or circuit breakers
6. Relays
7. Contactors
8. Magnetic starters

Figure 9–1 illustrates the electric power control system, showing the relationships of the parts.

The main disconnect and unit disconnect are similar in construction and function, although they are located in different parts of the system. Fuses and circuit breakers are similar wherever they are found in the system. Each of the electric control devices is sized for the electric current it must carry. Relays, contactors, and starters are electric switches in which one electric circuit controls another circuit.

9.3

Service Procedures for Electric Power Supplies

If trouble is suspected in the electric power controls on an air conditioning or a heat pump system, you, as a technician, must first determine whether there is electric power available in the system. If electric power is available, you must then determine whether it is the correct voltage. To make these checks, you should start at the main electric panel where power comes into the building.

To check the electric power supply to the building, open the main disconnect panel door and check the main disconnects or circuit breakers. These devices will be labeled "main disconnects" in the box. If the disconnects are closed, as indicated by the position of the handles, then there should be power to the branch circuits through the box.

To check the voltage coming into the panel, remove the inside cover of the panel and use a voltmeter to check across the terminals where the wires come into the box from outside the building. Be sure to use a voltmeter with a high enough voltage rating to cover the highest possible voltage available. If necessary, use high-voltage probes. Incidentally, the wiring on the input side of the panel or of any other device is calleld the **line side wiring**. The wiring leaving the panel or device is called the **load side wiring**.

To make the voltage check, set the voltmeter to a higher voltage than is expected in the supply system. Place one probe of the voltmeter on one of the wiring terminals on the line side of the main disconnect, and

Line side wiring
Load side wiring

place the other probe on another terminal. There are usually three incoming wires and three terminals. Place the probes on the first and second terminals and observe the voltage. Next, place the probes on the first and third terminals and observe the voltage. Finally, place the probes on the second and third terminals and observe the voltage. If the voltage readings are the same for all combinations of probe arrangements, then the system is a **three-phase system**. If a certain voltage is indicated between any two wires, and approximately half that voltage is indicated between either of those wires and the third wire, then the system is a **single-phase system**. (The Greek letter phi, ϕ, indicates the term *phase*.)

Three-phase system

Single-phase system

The main power to the building may be three-phase power, and the power in the branch circuits may be single-phase power. In this case, the branch circuit wiring is connected inside the panel to provide the single-phase power. There is nothing wrong with this arrangement.

The actual service voltage on a system varies according to the way the utility company takes the power from the main transformers in its lines. The electric utility company should be contacted to determine the voltage and phase for a specific job. Service procedures for checking single-phase and three-phase power supplies are discussed in the following subsections.

Single-Phase Power Supplies

The standard power supply for smaller air conditioning or heat pump systems is 230- or 240-volt, single-phase, 60 hertz. This power is shown on equipment tags as 230/240/1/60 power. This power can be furnished with either a two-wire system or a three-wire system. Figure 9–2 shows the two types of systems. The two **hot wires**— that is, the wires that carry current—in a single-phase circuit each have 115 or 120 volts potential to ground. They have 230 or 240 volts potential between them.

Hot wires

To check the voltage on a single-phase, 230-volt service, check the voltage between the two hot lines on the line side of the panel. Use a voltmeter, and set the meter for alternating current (AC) power and for the proper voltage scale. Place one of the probes of the volt-

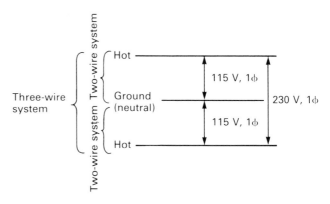

FIGURE 9–2
Schematic diagram of 115-volt and 230-volt electric power supply

Ground wire

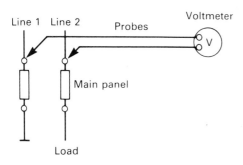

FIGURE 9–3
Checking voltage on single-phase system

meter on one of the power terminals where it enters the switch. Set the other probe on the other terminal, as shown in Figure 9–3.

The voltage reading on the meter should be within ±10% of the rated voltage of the system. If the meter reading is within the allowable range, then the electric power supply is all right. If the reading is not within the allowable range for the system, the power company should be notified.

Three-Phase Power Supplies

In systems using air conditioning or heat pump equipment with motors larger than 5 horsepower, three-phase electric power is commonly used rather than single-phase power. Three-phase electric power systems are either three-wire or four-wire systems. In a three-wire system, all three of the wires carry power. In a four-wire system, three of the wires are hot and the fourth wire is neutral. The fourth wire is a **ground wire** and does not carry current. A three-phase system has a rated voltage between any two hot wires and approximately half that voltage between a hot wire and the neutral. A disconnect switch for a three-phase power system should always have three sets of contacts on it. All three hot legs should be broken when the switch is opened.

To check the voltage on a three-phase system, use a voltmeter, as indicated in Figure 9–4. Check across two of the terminals on the line side of the disconnect. The line side is the side where the wires go into the disconnect box. The voltage reading between any two of the three hot wires on the load side of the disconnect should be the rated voltage of the system. If the voltage reading between any two of the hot wires varies more than 10% from the rated voltage for the system, the electric utility company should be notified.

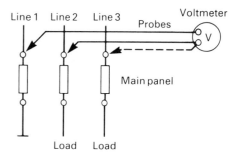

FIGURE 9–4
Checking voltage on three-phase system

9.4

Service Procedures for Electric Controls

If the voltage readings entering the main disconnect switch indicate the proper voltage for the system, the next checks to make are of the control devices. This section discusses the service procedures for checking the various controls. We begin with the disconnects.

Main and Branch Disconnects

The **main disconnect switch** for a system is a switch that allows an operator to turn off (or on) the electric power to the entire building. A **branch disconnect** is a switch on an individual branch circuit going from the main panel. Figure 9–5 illustrates a main electric control panel. A circuit breaker is used as a main disconnect switch and as overcurrent protection for the main circuit. Circuit breakers are also used as switches and overcurrent protection for each branch.

Main disconnect switch

Branch disconnect

A disconnect switch may be a knife-blade switch, or it may be a part of a circuit breaker. A typical knife-blade switch used as a main disconnect may or may not have fuses with the switch. A **knife-blade switch** has bladelike levers that pivot at one end and fit into contact points on the other end. A handle is used to open or close the levers. The electric circuit to be controlled is wired through the switch by connections at the terminal points. A typical knife-blade switch is shown in Figure 9–6. The switch is enclosed in a metal box, and the

Knife-blade switch

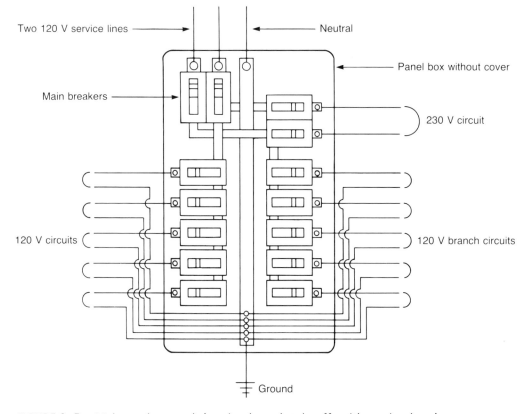

Two 120 V service lines — Neutral

Panel box without cover

Main breakers

230 V circuit

120 V circuits

120 V branch circuits

Ground

FIGURE 9–5   Main service panel showing branch takeoffs with service breakers

handle for operating the switch is located so that the switch can be opened or closed from outside the box. Disconnect switches are built for both two-wire systems and three-wire systems, which means that they are available as two-pole or three-pole disconnects.

To service either a main or a branch circuit disconnect switch, make a visual inspection first. Usually, you can see if the blade is opening and closing when the handle is moved. A knife-blade switch is a manual device, and so there is not much that can go wrong with it. In most disconnect boxes, the switch mechanism is covered by the front panel of the box. This panel may have to be removed in order for you to observe the switch action.

In a two-wire, single-phase system, such as the system used on a typical small air conditioning or heat

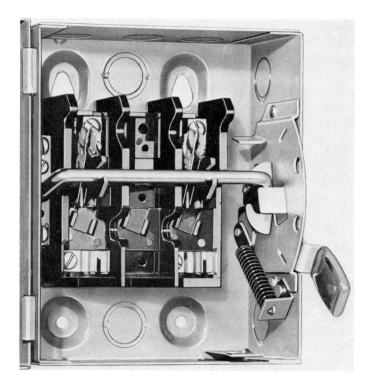

FIGURE 9–6
Knife-blade disconnect switch (Courtesy of Furnas Electric Co.)

pump system, both of the wires are hot. They must both
be opened by the disconnect switch when the power is
turned off. In a three-wire, single-phase system, a third
wire, or **neutral wire**, is used. This wire is connected to
a ground at the panel, and it does not carry current.
This wire does not have to be broken by the disconnect
switch. A two-pole disconnect switch can be used for
either a two-wire or a three-wire, single-phase system.

    If a visual inspection does not indicate anything
wrong with the disconnect switch, then the voltage
should be checked on the load side of the panel. To
check the voltage, follow the procedures described earlier for checking voltage on the line side of the system—
except that here, of course, you are checking on the load
side of the disconnect boxes.

    Safety note: When you are working on the load side
of the disconnects, make sure that the disconnect switch

Neutral wire

is open, whenever possible. The switch will have to be closed when voltage is being checked, but all safety precautions should be observed whenever the system is being checked with the electric power on.

Main Fuses and Circuit Breakers

Overload

When an electric system is designed, each circuit and each component in that circuit are sized for the electric current, measured in amperes, that will flow through the circuit or component during normal operation, plus 25% for a safety allowance. If an **overload** occurs in the circuit—that is, more electric current flows in a circuit than the wires or control devices are sized for—the wires and components overheat. Overheating is dangerous because of the fire potential.

Fuses
Circuit breakers

So that overload is prevented, fuses or circuit breakers are used in the circuit. **Fuses** or **circuit breakers** open the circuit in case of a current overload. The fuses or circuit breakers in a system that a service technician should first be concerned about are located in or adjacent to the main disconnect box. Figure 9–7 illustrates the fusing in a typical circuit for an air conditioning or a heat pump system. Service checks for fuses and circuit breakers are described in the following subsections.

Fuses. Fuses come in a variety of types, sizes, and amperage ratings. Two common types are cartridge and socket fuses. The type most often used for air conditioning or heat pump systems is the cartridge fuse. A

Cartridge fuse

cartridge fuse is a hollow fiber or plastic cylinder with a fuse wire running through it, as shown in Figure 9–8. The ends of the fuse wire are connected to metal caps on each end of the fuse. When the fuse is installed in a fuse holder, the two ends of the fuse fit into clips on the fuse holder that are connected into the circuit.

A fuse is used in each line that carries current in a circuit. If an overload occurs in the circuit, the fuse wire inside the cartridge melts, and the circuit is broken.

Socket, or screw-in, fuses are found in some circuits. These fuses are normally used with equipment that has a smaller size than the equipment the cartridge

Socket fuse

fuse is used with. A **socket fuse** screws into a receptacle,

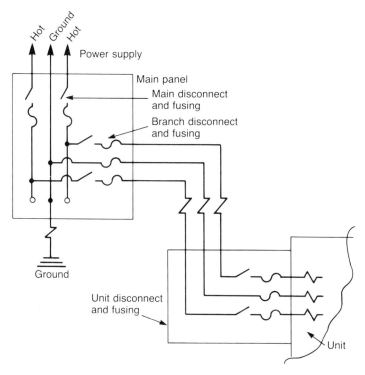

FIGURE 9–7
Wiring diagram showing fusing for typical air conditioning or heat pump system

FIGURE 9–8
Cartridge fuse

just as a light socket does. The terminals on the socket are wired into the electric circuit so that the current in the circuit flows through a fusible wire in the fuse. The fusible wire is sized so that it will melt if the current exceeds the rated amperage for the fuse.

A second type of socket fuse has a time-delay feature built into it. The **time delay fuse** does not open immediately if an overload occurs; rather, it allows current to flow for a short time before it opens. This fuse is used when a momentary overload is expected on start-up of a piece of equipment, but protection is still provided during normal operation. As shown in Figure 9–9, the time delay fuse includes a spring, a flexible connector, a screw base, and a contact.

In the checking of an electric circuit, a voltage check may show that there is electric power on the line side

Time delay fuse

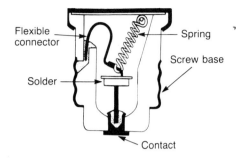

Flexible connector

Solder

Spring

Screw base

Contact

FIGURE 9–9
Internal parts of time-delay fuse

Continuity
Infinite resistance

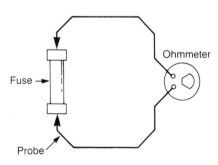

Fuse →

Ohmmeter

Probe

FIGURE 9–10
Fuse check with ohmmeter

of a fused disconnect box but no power on the load side. If the switch is closed, then the fuses should be checked to make sure that none of them are blown. In a two-wire, 230- or 240-volt system, there is a fuse in each of the lines. In a three-wire, 230- or 240-volt system, there is a fuse in each of the two hot lines but none in the neutral line.

There are two easy ways to check cartridge fuses in a single-phase system to see whether they are blown. One method is to take the fuses out of the fuse block and check for continuity with an ohmmeter. The other method is to check the fuses in place with a voltmeter. Each method is described next.

To check fuses with an ohmmeter, turn the power off by opening the disconnect switch. Remove the fuses from the fuse block with a fuse puller. As shown in Figure 9–10, place the two leads of the ohmmeter on the opposite ends of the fuse. If the meter reads zero resistance, showing **continuity**, the fuse is good. If the meter needle does not move, indicating **infinite resistance**, the fuse is burned out and must be replaced. Repeat this procedure for each of the fuses.

To check fuses left in place, use a voltmeter, and leave the electric power on. Make sure that there is power on the line side of the fuse holder. As shown in Figure 9–11, put one lead of the voltmeter on the line side of fuse 1, and put the other lead of the voltmeter on the load side of fuse 2. If there is no indication of voltage, then fuse 2 is faulty and should be replaced with a good fuse. If the voltmeter reading shows the nominal voltage for the system, then the fuse is good. To check fuse 1, place one lead of the voltmeter on the line side of fuse 2, and place the second lead on the load side of fuse 1. A no-voltage reading indicates that fuse 1 is bad. A nominal voltage reading indicates that it is good.

For a three-phase system, use the same method for checking fuses that you used for a single-phase system. However, there will be a third fuse to check. To check the fuses, use a voltmeter. With the probes, go from the line side of fuse 1 to the load side of each of the other two fuses, in turn. Then, go from the line side of either fuse 2 or 3 to the load side of fuse 1 to check fuse 1.

If the system being checked has screw-in fuses, you need only make a visual inspection to see whether the fuses are good. If a fuse is blown, the isinglass window in the top of the fuse will be discolored on the inside. If a fuse is burned out, replace it with a good one of the same amperage rating.

Circuit Breakers. Circuit breakers are mechanical devices that serve the same purpose in an electric circuit as fuses. They protect against current overloads. They also have a switch built into them so that they can serve as disconnects as well as fuses. Many smaller air conditioning or heat pump units have circuit breakers as combination disconnect switch and fusing.

Circuit breakers have an automatic time delay feature so that they will not open in case of a momentary overload, but they will function if the overload lasts longer than a few seconds. A typical circuit breaker is shown in Figure 9–12. Circuit breakers are wired in series in a circuit so that the current passes through them. A circuit breaker is used in each hot leg of a circuit.

Newer systems are often protected by circuit breakers. For circuit breakers used as overcurrent protection, an open is indicated when the lever on the breaker has moved to a position where a colored indicator is exposed. To reset the breaker, move the lever to the off position and then back to the on position. It is not necessary to replace the breaker unless it will not remain set when there is no extra electric load on the circuit.

Supply-Side Electric Wiring

The electric wiring that is external to the air conditioning or heat pump units—that is, the **supply side wiring**—is installed by a field electrician and is not really the responsibility of the troubleshooter or serviceperson. But problems caused by faulty wiring can affect the operation of the system. Thus, the power connections from the unit disconnect to the unit and the wiring of the controls that are external to the unit are the responsibilities of both the installer and the servicepeople.

Wiring within the unit is generally installed by the manufacturer of the equipment. However, all of the power wiring and control wiring may have to be checked when a problem in a system seems to originate in the

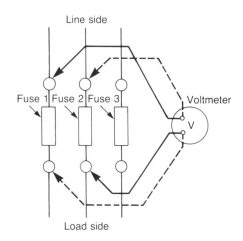

FIGURE 9–11
Fuse check with voltmeter

FIGURE 9–12
Typical circuit breaker (Courtesy of Furnas Electric Co.)

Supply side wiring

electric system. If there is power leaving the disconnect box and fuse panel, but the unit does not run, then the wiring in the system between the disconnect and the unit should be checked. Loose electric connections can cause the electric circuit to be broken entirely or to be broken intermittently. Or loose connections may simply cause an increase in amperage because of the added resistance in the connection.

A visual inspection of the wiring and the terminal connectors will usually suffice to show whether there is a problem in the wiring. The power should always be turned off while any work is being done on the wiring or connectors. To check the wire, look for places where the insulation has been scraped off, for burned insulation on the wire itself, or for burned spots on the wiring connections. Loose connections will normally show up as burned terminals or wires at the point of connection. If the inspections show that there is a problem in the wiring from the main disconnect to the unit disconnect, a qualified electrician should be called in to make the repairs.

If loose terminals are found, they should be tightened. If there are signs of burning at the connectors or on the insulation of the wires at the point of connection, the terminals and connectors—and possibly the wire—should be replaced.

On the low-voltage or control wiring, discoloration or burning will not always be present at the site of a loose connection. Erratic operation of the operating controls may be the only indication that something is wrong in the power circuits. If loose connections are suspected on the low-voltage circuits, the power should be shut off in the circuits that power the control circuits. Then, all the connections in the circuit should be tightened.

Unit Disconnects, Fusing, and Wiring

In addition to the main disconnect, there may be sub-panels located in the electric power distribution system. The **subpanels** are disconnect and fuse boxes located at strategic positions in the building, that is, positions close to the part of the system that they serve. For instance, a subpanel may be located near the air conditioning or heat pump units, and it is used to distribute electricity

Subpanels

to those units only. Other subpanels are used to furnish electricity to other equipment or to furnish lighting for the building. The subpanels are located between the main disconnect panel and the unit disconnect box in the wiring system.

Besides the main disconnect or subpanel disconnect, an air conditioning or a heat pump system normally has disconnects and possibly fusing at each of the units in the system. **Unit disconnects** usually are internally fused.

Unit disconnects

Figure 9–13 shows a typical wiring arrangement for a rooftop system. The subpanel is located between the power source and the unit disconnect. The wiring from the unit disconnect runs to the control devices inside the unit. To check the unit disconnect, fusing, and wiring, use the same procedures as used for the main disconnect, fusing, and wiring.

Magnetic Relays

Inside the unit, the electric power circuit goes from the unit disconnect to the various electric devices, such as motors, through a control device. Depending on the

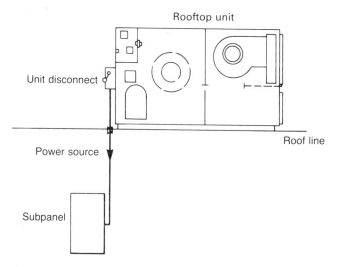

FIGURE 9–13
Wiring diagram of branch circuit showing subpanel and unit disconnect

Magnetic relay

Electromagnetic coil

FIGURE 9–14
Magnetic relay (Courtesy of Furnas
Electric Co.)

electric load, this control device can be a relay, a contactor, or a magnetic starter. In smaller units, a magnetic relay is used. A **magnetic relay** is an electric switch in which one electric circuit is used to control another circuit. A typical magnetic relay is shown in Figure 9–14. In most cases, a relay is used to control a power circuit with a control circuit.

The wiring symbol for a magnetic relay is shown in Figure 9–15A. The relay has an **electromagnetic coil** that pulls in an armature, as shown in Figure 9–15B. The armature has a set of contacts. These contacts match another set on the frame of the relay. The coil and contacts are wired in series in the control circuit. As the armature moves back and forth when the control circuit is energized or de-energized, the contacts on the armature open or close. As the contacts close, the circuit is made. And as the contacts open, the circuit is broken.

If the electric system has power leaving the unit disconnect box, but the electric components of the unit still do not operate, then the control relay should be checked. To check a magnetic relay to see whether it is functioning properly, check both the magnetic coil and the contacts.

To check a magnetic relay coil, first turn off the power to the unit at the unit disconnect. Then, disconnect the relay coil from the circuit by removing the control circuit leads from the coil terminals. Using an

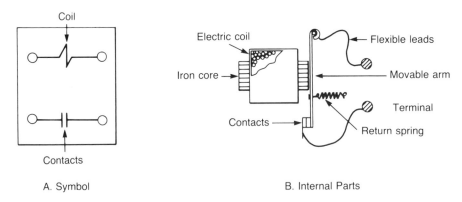

A. Symbol   B. Internal Parts

FIGURE 9–15   Wiring symbol and parts for magnetic relay

ohmmeter, check across the coil for continuity. Place one probe of the ohmmeter on one terminal of the coil, and place the other probe on the other terminal, as shown in Figure 9–16. Observe the meter needle. Deflection of the needle indicates continuity, that is, indicates that the coil is good. No deflection of the needle indicates that the coil has an open and is not good. In this case, the coil should be replaced.

To check the relay contacts, make a visual check of the armature and contacts first. See whether the armature can be moved manually. Then, inspect the contact points to see whether they are burned or welded together. If the armature cannot be moved, the contacts probably are welded together. If the contacts are pitted, burned, or welded together, they should be replaced. Next, close the disconnect switch and energize the relay coil by turning the thermostat to call for cooling. If the contacts are visible, observe whether or not they close. If they do not close when the coil is energized, then the relay coil is bad and should be replaced.

If the contacts are not visible on the relay, energize the coil, and use a voltmeter to check across the contacts for an open. As shown in Figure 9–17, place one probe of the meter on the terminal of one side of the contacts, and place the other probe on the terminal of the other side. Observe the voltmeter needle. If it shows a voltage reading, indicating an open, then the relay is bad and should be replaced. If the voltmeter shows no voltage, then the contacts are closed, and the relay coil is good.

Contactors

A **contactor** is an electric device similar to a relay. It is used in circuits where relatively high amperage is expected. Thus a contactor is usually a heavy-duty relay. A contactor has an electromagnetic coil that is energized by the control circuit. The magnetic field of the coil moves an armature back and forth. The armature has a set of contacts, and a matching set of contacts is mounted on the frame of the contactor. As the armature moves, the contacts open or close. The contacts are wired into the power circuit.

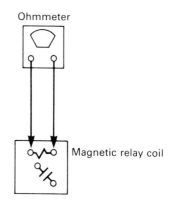

FIGURE 9–16
Checking magnetic relay coil

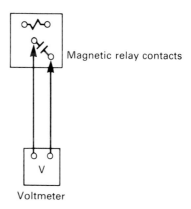

FIGURE 9–17
Checking magnetic relay contacts

Contactor

FIGURE 9–18
Contactor (Courtesy of Furnas Electric Co.)

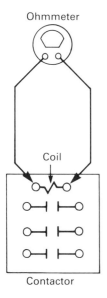

FIGURE 9–19
Checking contactor coil

A typical contactor is shown in Figure 9–18. Contactors are often used as motor controls as well as unit control devices. There may be several contactors on a system, depending on what they are used to control.

The procedure for checking a contactor is nearly the same as the procedure for checking a relay. First, check the contactor coil by turning off the power and disconnecting the leads from the coil. Then, use an ohmmeter to check the coil for continuity by placing one probe on each end of the coil wire, as shown in Figure 9–19. If there is continuity, as shown when the needle indicates no resistance, then the coil is good. If the meter needle indicates no continuity by showing infinite resistance, then there is an open in the coil, and it should be replaced.

Next, check the contacts. Make sure that the power is turned off to the unit, and make a visual inspection. If the contacts are pitted or discolored, or if they are welded shut, clean or replace the contacts or the entire contactor. Turn the power on and energize the coil. Visually observe whether the contacts are pulling in. If they are not, then check to see whether there is some physical obstruction preventing movement. If there is no physical obstruction, the contactor is bad and should be replaced.

Note: Some contactors are position-sensitive and require the help of gravity for movement of the armature. Make sure that the replacement contactor is installed in the same position on a unit as the contactor being replaced.

As a final check, use a voltmeter to check the contacts. With the contactor energized, place one probe of the meter on the line side of one set of contacts, and place the other probe on the load side of the same set, as indicated in Figure 9–20. Observe the meter. A no-voltage reading means that the contacts are closed and that the contactor is good. A voltage reading indicates that the contacts did not close and that the contactor is bad. It should be replaced.

Repeat the same check on each set of contacts to determine whether all of the contacts are operating as they should. If a three-phase contactor is being checked, there will be three sets of contacts. Use the same procedure for each set of contacts.

Magnetic Starters

A **magnetic starter**, such as the one shown in Figure 9–21, is used as a motor control device more often than it is used as a unit control device. It is similar in construction to a contactor, but with one notable exception. A magnetic starter has overload protection built into it for the circuit that it controls.

The holding coil, or magnetic coil, circuit has **overload heaters** wired into it, as shown in Figure 9–22. These heaters open the contacts if the amperage on the circuit exceeds a set limit. Magnetic starters are used in the same way as contactors, except that they are used to control motors that do not have inherent overload protection.

To check a magnetic starter for proper operation, use the same checks for the coil and the contactors as used for those parts on a contactor. (See the preceding section for the procedure.) The heaters can be checked by disconnecting the wires and using an ohmmeter to check the heaters for continuity. If they are burned out, an open will be indicated. If the heaters are burned out, they should be replaced. Be sure to find the cause of the burnout, and make corrections before the power is turned back on.

9.5

Summary

Most modern air conditioning or heat pump systems use electricity as a source of power to operate the equipment in the system and to control the operation of the equipment. The main parts of the electric power supply system are the main disconnects, the unit disconnects, the wiring, the main and the unit fuses or circuit breakers, the relays, the contactors, and the magnetic starters. Both single-phase and three-phase electrical power are used in air conditioning and heat pump systems. Small units usually use single-phase power, and large units use three-phase power. Two-wire, three-wire, and four-wire electrical distribution systems are used, depending on the equipment to be powered.

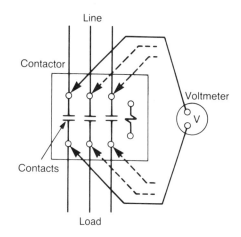

FIGURE 9–20
Checking contactor contacts

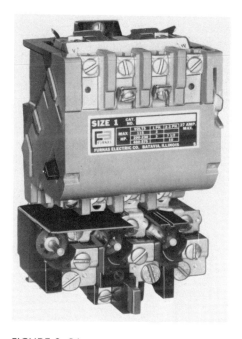

FIGURE 9–21
Magnetic starter (Courtesy of Furnas Electric Co.)

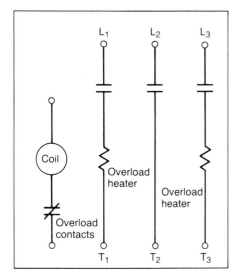

L_1 L_2 L_3

Coil

Overload heater

Overload heater

Overload contacts

T_1 T_2 T_3

FIGURE 9–22
Wiring diagram of magnetic starter showing heaters

There are specific steps you must take when checking each of the various parts of the electric power control system. Electric connections and wiring can be checked by visual inspection, but many of the electric parts must be checked with meters. An ohmmeter is used to check for continuity through circuits or parts, with the electric power off. A voltmeter is used to check for electric potential, with the electric power on.

Since the electric power system is an important part of the total air conditioning or heat pump system, a service technician must be familiar with the system. The technician must be able to check the electric control components to determine whether they are causing problems in the system.

9.6

Questions

1. Name the three main parts of an electric power system.
2. Name the eight common controls in a power system.
3. What is the first service check to make if trouble is suspected in the electric power supply system?
4. Define the terms *line side* and *load side*.
5. Is the following statement true or false? A single-phase power system always has only two wires.
6. Is the following statement true or false? A three-phase power system has two hot wires and a ground wire.
7. Describe the part of the electric power system that is controlled by the main disconnect. Describe the part that is controlled by the branch disconnect.
8. What is the purpose of a fuse as it is used in an electric power system?
9. Describe the functions of a circuit breaker.
10. What does the term *continuity* mean?
11. Is the following statement true or false? To check fuses with an ohmmeter, you must take them out of the circuit.
12. A magnetic relay is an _____ switch in which one circuit controls another.

13. Name two things that must be checked if a magnetic relay is not functioning.
14. Describe a contactor.
15. How does a magnetic starter differ from a contactor?

10

Control System

Control system

10.1

Introduction

The operation of an air conditioning or a heat pump system depends on the controls in the system. The **control system** regulates the operation of the individual parts of the system so that they work harmoniously to produce the desired results. Just as there are many different control functions in a system, there are many controls also. A malfunction of one part of the control system affects the entire system just as much as a malfunction of a mechanical part would.

Before a serviceperson can determine whether a control system is operating properly, he or she must know how that control system operates normally. The technician must also know how its operation affects the total system.

This chapter describes the operation of the control system for a typical air conditioning or heat pump system. Also, the use of each control is explained. Finally, procedures are outlined for checking the controls for failure when they are suspected of being faulty.

10.2

Types of Control Systems

There are three main types of control systems used in the majority of air conditioning and heat pump systems: electric, pneumatic, and electronic. Most of the small comfort systems use **electric controls**. Systems in large buildings often use **pneumatic controls**, but many of the individual controls in the equipment are electric. **Electronic controls** are becoming more common, and some of the controls used in the equipment being manufactured today are electronic controls.

Electric controls
Pneumatic controls
Electronic controls

Since most of the controls used at the present time on typical small systems are electric controls, they are the only controls that are covered in detail in this text.

10.3

Function of the Control System

The **function** of an air conditioning or a heat pump system is to produce a cooling or heating effect at a fixed rate, even though the cooling or heating load requirements vary. Thus, it is necessary to control the operation of the unit so that it will produce the average amount of cooling or heating required over a given time period, to match the load. This control is achieved by cycling the unit on and off.

Function of system

A thermostat in the building being cooled or heated senses the temperature of the air and turns the cooling system on if the temperature in the building goes above the temperature at which the thermostat is set, that is, the **set point** of the thermostat. The thermostat also turns the unit off if the temperature in the building goes below the set point of the thermostat. The thermostat controls the on/off action of the equipment by controlling an electric control circuit. Other control devices are used to operate individual components in the system.

Set point

10.4

Control Panel

A **control panel** is located inside the air conditioning or heat pump unit cabinet. This panel has terminal blocks for electric power connections and for control circuit connections. It also contains the various controls that operate the parts of the system. A schematic wiring diagram

Control panel

of the unit control system is fastened inside this panel. The various controls are identified on the diagram.

10.5

Operating Controls

There are usually many controls in an air conditioning or a heat pump system. Different controls have different functions. To make it more convenient to describe them, and to explain the service procedures for each, we will divide the controls into two categories, according to their function in the system. The two categories are operating controls and safety controls.

The operating and safety controls for a typical air conditioning or heat pump system are illustrated in Figure 10–1. The operating controls include the thermostat, fan switch, compressor relay coil, evaporator blower relay coil, transformer, evaporator blower relay contacts, and compressor relay contacts. The safety controls include the high-pressure switch and the low-pressure switch. The operating controls are discussed in this section; safety controls are discussed in Section 10.6.

Operating controls are the controls that turn the unit off and on and also operate individual components in the system, such as motors. In many cases, the operating controls are relay controls. As defined in Chapter 9, a relay control is a device that controls one electric circuit with another circuit. The circuit doing the controlling is called the **control circuit**, and the circuit being controlled is called the **power circuit**. In most air conditioning or heat pump systems, the control circuit is a low-voltage circuit, while the power circuit voltage is the rated voltage of the system's equipment.

In a typical air conditioning or heat pump system, the operating control system includes the transformer, thermostat, control relays, and starting relays of various types. Some of the other electric devices used in the system, such as capacitors, may be called controls, but their function in the system is one of regulation rather than control. This chapter deals with devices that have

Operating controls

Control circuit
Power circuit

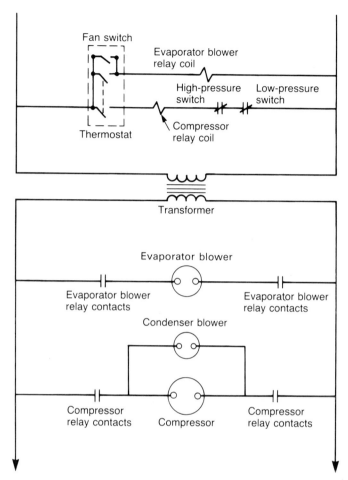

FIGURE 10–1

Operating and safety controls on typical air conditioning or heat pump system

a control function; the other devices are covered in other chapters in the text.

Figure 10–2 shows the operating controls in a typical air conditioning or heat pump system. Descriptions of each of these controls, of their use in a system, and of the method used for troubleshooting them are given in the following subsections.

Transformer

Primary winding

Secondary winding

Step-up transformer

Step-down transformer

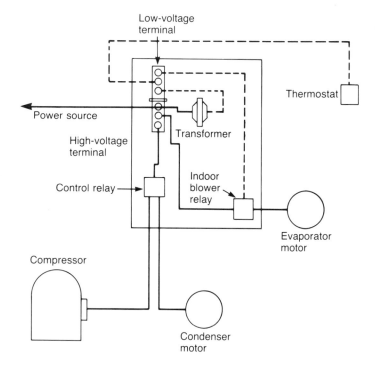

FIGURE 10–2
Operating controls on air conditioning system

Transformer

A **transformer** is an electric device that changes the voltage in an alternating current electric system from one voltage to another. A control transformer is shown in Figure 10–3. A transformer can be used to either increase or decrease the voltage.

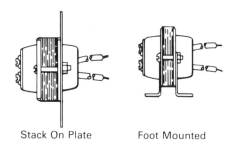

Stack On Plate   Foot Mounted

FIGURE 10–3
Low-voltage control transformer

Description. Physically, a transformer has two magnetic coils wound on the same core, as shown in Figure 10–4. One set of windings, called the **primary winding**, is connected to the power source. The other set of windings, called the **secondary winding**, is connected to the load. A magnetic field generated by the electric current in the primary coil generates a current in the secondary coil by induction. If the primary coil has a greater number of windings than the secondary coil, then the voltage in the secondary coil will be less than the voltage in the primary coil. If the primary coil has fewer windings than the secondary coil, then the voltage in the secondary coil will be higher than the voltage in the primary coil.

A transformer that has a higher voltage in the secondary coil than in the primary coil is called a **step-up transformer.** A transformer that has less voltage in the secondary coil than in the primary coil is called a **step-down transformer**. A step-down transformer is used in most control systems to provide lower-voltage electric power for the control system. Low voltage in the control system allows the use of smaller wire for the control circuits than can be used in the power circuits and the use of lighter controls in general. One of the biggest advantages of low voltage is that the switches and contacts in the sensing elements are light, making the controls more sensitive.

The transformer in a control circuit is connected so that the primary side of the transformer is wired to the line voltage source. The secondary side is connected to the control circuit.

Servicing. To check a transformer for proper operation, first check the voltage to the primary side of the transformer. The primary voltage should be within ±10% of the primary voltage rating for the transformer. If the primary voltage varies from the rated voltage, the secondary voltage will vary also. To check the voltage, set a voltmeter to the expected voltage of the power circuit. Place one of the meter probes on one terminal of the transformer primary coil, and place the other probe on the other terminal, as shown in Figure 10–5. Read the voltage on the meter scale. If the voltage is not within the allowable range for the transformer, then the transformer should be changed. The transformer's voltage, phase, and hertz are shown on its nameplate.

If the meter indicates that the voltage is within the allowable range, then the secondary voltage should be checked. This voltage is checked in the same way as the primary voltage is checked, except that the meter is set for the output of the transformer and the probes are placed on the terminals of the secondary coil. If there is no indication of voltage on the secondary side of the transformer, then the transformer is faulty.

To double-check a transformer, turn off the power

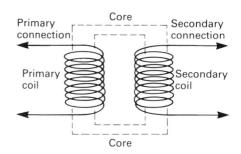

FIGURE 10–4
Primary and secondary coils in transformer

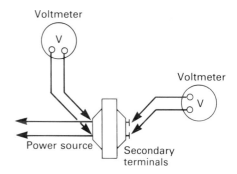

FIGURE 10–5
Checking voltage to transformer

Thermostat

Bimetal element

Mercury bulb switch

Anticipator

and disconnect the transformer from the circuits on both sides of the system. Use an ohmmeter and check for continuity through each coil. For the continuity check, place one probe of the ohmmeter on each terminal of the primary coil. If the meter needle shows infinite resistance, the coil is burned out. Repeat the check for the secondary coil. A burnout on either coil means that the transformer will have to be replaced.

Thermostats

Most air conditioning and heat pump systems are controlled by a thermostat. The **thermostat** turns the unit on when a need for cooling or heating arises, and it turns the unit off when no cooling or heating is needed. A cooling-only system has a cooling-only thermostat. A heat pump or a heating-and-cooling system has a combination heating/cooling thermostat. A typical cooling thermostat, such as the one shown in Figure 10–6, is a switch that closes when the temperature of the material being sensed rises. The switch opens when the temperature drops. A heating thermostat has a switch that opens on temperature rise and closes on temperature drop.

Description. In most thermostats the switch is activated by a bimetal element. A **bimetal element** is made up of two pieces of metal that expand and contract at different rates when the temperature changes. Thin strips of the two metals are fastened together to form the element. When the temperature increases, one of them expands more than the other, and the element bends in the opposite direction. If the temperature decreases, the element bends in the other direction. In most thermostats, this element is formed into a coil. If the temperature increases, the coil winds in one direction. If the temperature decreases, the coil unwinds in the other direction. One end of the bimetal is fixed in place, and the other end is free to move. The switch is attached to the movable end. As the bimetal winds or unwinds, the switch opens or closes.

Figure 10–7 shows the wiring and connections for

FIGURE 10–6
Typical cooling thermostat (Courtesy of Honeywell Inc.)

a thermostat. The cooling anticipator labeled in Figure 10–7 is described in a later subsection.

The most common type of thermostat uses a mercury bulb switch as the electric switch. The **mercury bulb switch**, which is also shown in Figure 10–7, is a closed glass tube with a drop of mercury inside. The tube has wire contacts fused into it at one end. The wires are connected to terminals with flexible connections, as indicated in Figure 10–8. The bulb is mounted on the bimetal element in such a way that when the element moves in response to temperature changes, the bulb is tipped back and forth. As the bulb tips one way or the other, the mercury runs from one end of the glass bulb to the other. When the mercury runs to the end with the terminal wires, the mercury, which is a conductor, completes the electric circuit to which the wires are connected. When the bulb tips the other way, the mercury runs to the end that has no wires, and the circuit is broken. The thermostat is calibrated so that the bulb will tip enough to allow the mercury to run from one end to the other with only a few degrees of temperature change. This calibration provides the differential needed to prevent the unit from short cycling.

The bimetal element for the mercury bulb switch is mounted inside the thermostat on a backplate that is movable. The backplate is attached to the set point indicator of the thermostat. When the set point indicator is moved, the make/break point of the thermostat switch coincides with the set point.

Anticipator. The sensitivity of a thermostat in reacting to room temperature is increased if an anticipator is built into it. The anticipator in a cooling thermostat is a small electric heater inside the thermostat case that provides heat during an off cycle. Figure 10–9 illustrates an anticipator in a cooling thermostat. The heat provided by the anticipator causes the thermostat to call for cooling sooner than it would if it only sensed room temperature. The anticipator heater is wired in parallel with the bimetal switch. This connection causes the electric current from the control circuit to go through the heater during an off cycle. But current bypasses the heater when the switch is closed, because there is less resistance through the switch contacts than through the heater. The heater

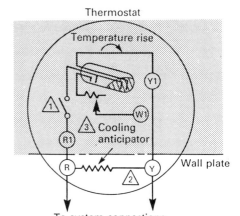

To system connections
COOLING ONLY THERMOSTAT

1. Model with positive off switch opens the thermostat circuit when set point dial is moved to the off position.
2. Make system wiring connections to terminals on wall plate.
3. R1, Y1 terminals on thermostat are directly connected to R, Y terminals on wall plate when thermostat is mounted on wall plate.

FIGURE 10–7
Schematic diagram of wiring inside typical thermostat (Courtesy of Honeywell Inc.)

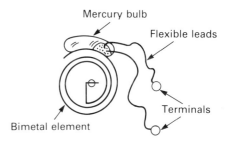

FIGURE 10–8
Mercury bulb switch for thermostat

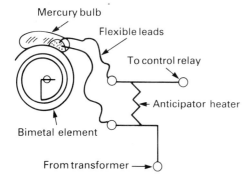

FIGURE 10–9
Anticipator in cooling thermostat

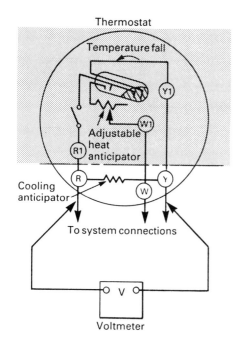

FIGURE 10–10
Checking thermostat circuit with voltmeter

is sized to provide the amount of heat needed for the proper anticipation in relation to the current draw of the control circuit.

A heating thermostat also has an anticipator. But in the heating thermostat, the heater is only heated during an on cycle. This feature provides false heat when the heating unit is on and shortens the on-cycle time of the unit.

Servicing. When checking a thermostat for proper operation, a service technician should first check the set point of the thermostat. Because the people who control thermostat settings on most installations do not understand the function of the controls, the technician should personally check the setting. A visual check will show the setting.

If a thermostat is suspected of being out of calibration, the calibration should be checked. To check the calibration, compare the make/break set point of the thermostat with an accurate thermometer. If the temperature at which the thermostat switch opens and closes, as shown on the set point indicator, does not coincide with the actual temperature, the thermostat is out of calibration. Since the calibration is set at the factory, only a factory-trained serviceperson should attempt to recalibrate one internally. Usually, all that the technician is required to do is to make sure that the thermometer on the thermostat has the same reading as the set point indicator when the thermostat switch opens and closes.

Some combination heating/cooling thermostats have a system switch that allows an operator to choose the cooling mode or the heating mode. Others are automatic. When servicing an air conditioning control system, check the system switch on the thermostat to make sure that it is in the cooling position.

Suppose that the thermostat system switch is calling for cooling, and the set point of the thermostat is set low enough to call for cooling, but the unit is still not running. In this case, check the operation of the thermostat with a voltmeter. Place one of the leads of the meter on the low-voltage power lead coming from the transformer to the thermostat, as shown in Figure 10–10. This terminal is the R terminal. Place the other lead on the thermostat terminal that is connected to the unit control relay. This terminal is the Y terminal. If the meter indicates the proper control voltage, the thermostat is not working and should be replaced. If there is no voltage shown on the meter, the thermostat is working properly. The problem lies somewhere else in the control system.

Thermostat operation can also be checked with a jumper wire. If the unit is not running, but the thermostat is calling for cooling, a jumper wire can be placed across the input terminal and the output terminal of the thermostat. If the unit starts with the jumper in place, the thermostat is faulty.

Short cycling of an air conditioning or heat pump unit—that is, turning off and on in short periods of time—can be caused if the anticipator is set incorrectly. The anticipator setting determines the length of the off and on cycles of the unit. The anticipator should be set to match the amperage that flows through the thermostat control circuit.

To check the amperage draw in the thermostat circuit, use an ammeter. A regular clamp-on ammeter can be used. Since most ammeters do not read amperage in a low enough range for control circuit amperage, a multiplier is used. The **multiplier** is a coil of control wire with a known number of turns (usually, ten turns) in the coil. The coil is wired into the control circuit, and the ammeter is clamped around the coil of wire, as shown in Figure 10–11. The reading shown by the meter is actually multiplied by the number of turns of wire. So, you should divide the reading by the number of turns to find the actual amperage in the circuit. The anticipator in the thermostat is set for the amperage reading determined by this method.

Terminal Strip

A high-voltage terminal strip and a low-voltage terminal strip are located in the control panel on the unit. Line voltage from the unit disconnect runs to the **high-voltage terminal strip**. Power from the transformer runs to the **low-voltage terminal strip**. Line voltage power goes from the high-voltage terminal strip to the line voltage controls and devices in the system. Low-voltage power goes

Short cycling

Multiplier

High-voltage terminal strip

Low-voltage terminal strip

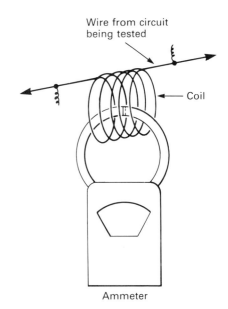

Wire from circuit being tested

Coil

Ammeter

FIGURE 10–11
Using coil of wire to multiply amperage readings

from the low-voltage terminal strip to the low-voltage controls in the system.

Control Relay

The main control located in the control panel is the control relay for the system. The **control relay** is either a magnetic relay or a contactor, depending on the size of the equipment being controlled. In either case, the control function is the same. The relay contacts are wired in series in the circuit from the high-voltage terminal strip to the various devices in the system that are energized by line voltage, such as the compressor. The relay coil is wired in series in the control circuit from the thermostat. This circuitry is shown in Figure 10–12 for a typical air conditioning system.

The most common type of relay used in control systems, either for control relays or for control of individual parts, is the magnetic relay. It is called a **magnetic relay** because it uses a magnetic coil to produce a mag-

Control relay

Magnetic relay

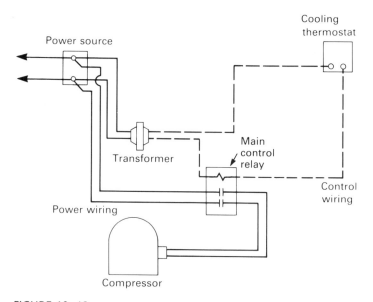

FIGURE 10–12
Wiring circuitry for main control relay in typical air conditioning system

netic field that actuates an armature that opens or closes an electric switch. A complete description of a magnetic relay and of service procedures used in checking relays is given in Chapter 9. Some special relays are used for starting switches on motors. These relays are described in Chapter 12.

Liquid Line Solenoid Valve

A **liquid line solenoid valve** is an electrically controlled flow valve. It is used to control the flow of the liquid refrigerant in the system. This valve is installed in the liquid line ahead of the expansion device in a system with a pump-down control. A **pump-down control system** is one in which the thermostat controls the flow of the refrigerant, and the low-pressure switch controls the operation of the compressor.

Liquid line solenoid valve

Pump-down control system

Description.   A solenoid valve is a positive-acting, on/off valve. When the thermostat calls for cooling, the valve opens. Then, liquid refrigerant flows through the valve to the expansion device and into the low-pressure side of the system. This flow causes an increase in the low side pressure, and the low-pressure switch turns the compressor on. When the thermostat is not calling for cooling, the valve is closed. Then, the pressure drops in the low-pressure side of the system, and the unit is shut down by the low-pressure switch.

A solenoid valve has a valve stem that opens or closes the valve port by moving up or down. A solenoid coil, which is an electromagnetic coil, surrounds the upper part of the valve stem, as shown in Figure 10–13. When the solenoid coil is energized, the stem of the valve is drawn up into the magnetic field. When the coil is de-energized, the weight of the stem and the refrigerant flow cause the valve to close.

Servicing.   To check a solenoid valve for proper operation, first check the electric connections to the valve. The valve coil is connected in the control circuit from the thermostat or other operating control for the system. Make sure that the terminal connections are tight.

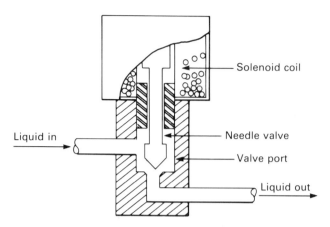

FIGURE 10–13
Internal parts of liquid line solenoid valve

Next, check the voltage on the circuit when the thermostat is calling for cooling. The voltage at the valve coil should be the rated voltage for the control system. Before making this check, make sure that the solenoid coil has the same voltage rating as the control system. To make the check, use a voltmeter, and set it on the appropriate scale for the voltage of the system. As indicated in Figure 10–14, place one of the meter probes on the terminal of one side of the solenoid coil on the valve. Place the other probe on the other terminal. Read the voltage on the meter. If the reading is within ±10% of the rated voltage of the system, but the valve does not open, then the solenoid coil is faulty.

To check the solenoid coil for an open in the winding, use an ohmmeter. Disconnect the control circuit wires from the coil terminals. Place one of the probes of the ohmmeter on the terminals on each side of the coil, and read the meter. If the meter indicates a very low ohm reading, showing continuity through the coil, the coil is good. If the meter indicates a very high ohm reading, showing no continuity, the coil has an open and should be replaced.

Another possible source of trouble with a solenoid valve is that the valve stem may stick. Since the valve stem has to be free to move when the coil is energized, any dirt or corrosion on the stem can affect it. A visual

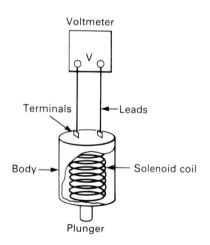

FIGURE 10–14
Checking solenoid coil with voltmeter

inspection should show if the valve stem is free to move. If any dirt or corrosion is found on the stem, the stem should be cleaned with emery cloth or steel wool. If the valve will be exposed to further contamination, then it should be shielded or otherwise protected.

10.6

Safety Controls

Safety controls are used on air conditioning and heat pump systems to protect the equipment from damage in case of a malfunction. Short cycling of a compressor can cause a burnout of the motor windings in the compressor. The cause of the short cycling may be fairly easy to detect and correct, but replacement of the compressor is a major project and is relatively expensive.

Figure 10–15 shows the safety controls for a typical air conditioning or heat pump system. Safety controls

Safety controls

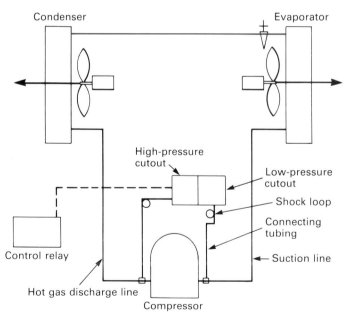

FIGURE 10–15
Safety controls for air conditioning or heat pump system

FIGURE 10–16
Dual-pressure switch (Courtesy of
Ranco Controls Div.)

Pressure switch

Low-pressure switch

High-pressure switch

are found in many parts of a control system. In most cases, safety controls are electric devices and are located in the electric circuits that control parts or operate the parts that the controls are designed to protect.

Some of the safety controls in a typical air conditioning or heat pump system are electric fuses, pressure switches, and motor overloads. Fuses were discussed in Chapter 9. In the following subsections, pressure switches and motor overloads are described, their functions in the system are explained, and service procedures for each are covered.

Pressure Switches

A **pressure switch** is a control that senses the pressure in some part of a system and operates an electric switch in response to that pressure. A **low-pressure switch** senses the refrigerant pressure in the low-pressure side of the system. A **high-pressure switch** senses the refrigerant pressure in the high-pressure side of the system. If the refrigerant pressure drops below or rises above a safe operating pressure, then the switches open and the compressor is turned off.

Description. A typical pressure switch used in an air conditioning or a heat pump system is shown in Figure 10–16. A pressure switch has a very small hollow tube that is connected to the refrigerant line in the part of the system where the pressure is to be sensed. The other end of the tube leads to a closed container that has a bellows or diaphragm element. Pressure in the tube causes the element to move. The amount of movement of the element is in direct proportion to any pressure change.

The element is connected by a mechanical linkage to a mechanism that opens or closes an electric switch. The electric switch is wired in series in the control circuit between the thermostat and the control relay, as shown in Figure 10–17. If the switch opens, the unit will shut off. The pressure switch has an adjustable set point and differential, and they can be adjusted for the low or high pressure at which the switch will cut out the operation of the compressor.

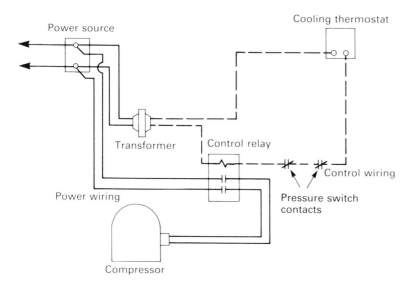

FIGURE 10–17 Pressure switch contacts in typical control system

Servicing. To troubleshoot a pressure switch, install a pressure gauge in the part of the system in which the pressure is being sensed. With the unit running, compare the pressure shown on the gauge with the pressure setting of the switch. Determine whether the switch should be open or closed. This contact depends on the pressure setting on the switch. If the switch does not open or close, change the pressure setting both above and below the actual pressure in the system. If the switch does not open and close when it should as the pressure changes, the switch is faulty and should be replaced.

Sometimes, the electric contacts in a pressure switch become welded shut, and hence, they will not open when they should. To check for welded contacts, turn off the electric power to the unit and disconnect the electric connections at the switch. Use an ohmmeter and check for continuity through the contacts in the switch. For this check, place one of the probes of the ohmmeter on the terminal on one side of the switch, and place the other probe on the other terminal, as indicated in Figure 10–18. An infinitely high resistance reading shows that the switch contacts are open. If the meter shows a very low resistance, then the contacts are closed.

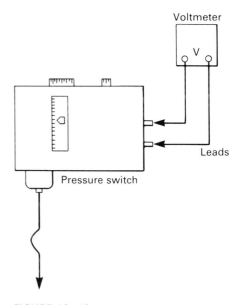

FIGURE 10–18
Checking operation of pressure switch with voltmeter

Motor overloads

Two-pole overload

Three-pole overload

Four-pole overload

In-winding thermostat

In large compressors, a pressure switch is often used in the lubricating oil line to make sure that the bearings in the compressor receive proper lubrication. An oil pressure switch works as the refrigerant pressure switches just described, except that the tube connecting the switch to the system goes to an oil line. Use the same procedures for checking an oil pressure switch as used for checking refrigerant pressure switches.

Motor Overloads

Motor overloads are switches that are operated by temperature or an electric current. The switch in an overload is wired into the control or power circuit of a motor. The overload protects the motor against damage if the temperature or amperage gets too high. When a motor is overloaded, the amperage increases. Increased amperage can cause the motor to get so hot that the insulation on the wires in the coils breaks down. When the insulation breaks down, an electric short occurs. If a high-temperature or a high-amperage condition exists, the overload switch opens, and the motor is shut off until the motor cools off.

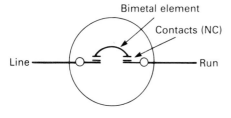

FIGURE 10–19
Two-pole motor overload

Description.   The simplest overload is the temperature-sensing overload. It has two electric terminals and is called a **two-pole overload**. The two-pole overload control is mounted on the motor or on the shell of a hermetic compressor, where it senses the temperature of the motor. It has a bimetal element, as illustrated in Figure 10–19, that opens and closes a switch, depending on the temperature it senses. The switch contacts are normally closed (NC). Sometimes, a two-pole overload is wired into the control circuit for the motor. At other times, it is wired into the power circuit.

A second type of overload is the **three-pole overload**. It has an electric heater inside that is wired in series with the motor power circuit. If an overcurrent condition exists, the heater gets hot and causes the bimetal to warp and open the switch. Figure 10–20 illustrates a three-pole overload. The three terminals are common, run, and start. A three-pole overload is used when the installer wants to sense the line current with the heater and also break the power leg to the motor in case of

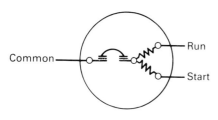

FIGURE 10–20
Three-pole motor overload

an overload condition. This overload control will also open if it senses a temperature rise of the surface on which it is mounted.

A third type of overload control is the **four-pole overload**. In this overload, the heater and the contacts are separate from each other. One set of terminals connects the ends of the heater, and another set connects the ends of the contactor, as shown in Figure 10–21. The four-pole overload is used when the heater must sense the load amperage, and the contacts are wired into the control circuit.

In some types of motors, an **in-winding thermostat** is used as an overload control. This device is actually a small thermostat that is embedded in the motor windings at the time the motor is built. Figure 10–22 illustrates an in-winding thermostat used for overload protection. The leads from the thermostat are connected in series in the motor control circuit. If the temperature of the motor windings goes above a predetermined setting, the thermostat contacts open, and the motor is shut off.

Servicing. If the motor overloads are suspected of being at fault on a troubleshooting or service call, they should be checked to see whether they are functioning properly. To check overloads, open the disconnect so that the motor circuit is de-energized. Then, disconnect the overload from the circuit it is in. Check the contacts by placing one lead of an ohmmeter on one terminal of the overload contact circuit. Place the other lead of the meter on the other terminal, as indicated in Figure 10–23. The meter should show continuity across the terminals any time that the overload is not overheated. If there is no continuity, the contacts in the overload are not opening. They are probably welded shut from an overcurrent condition. No continuity indicates a faulty overload, and it should be replaced.

The heater in a three-pole or a four-pole overload should also be checked to see whether it is burned out. An ohmmeter is used again, and the same method is used to check the heater coil as is used to check the contacts. If the heater has an open, indicating that it is faulty, then the meter will show no continuity through it. In this case, the overload should be replaced.

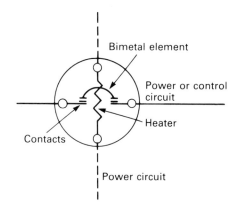

FIGURE 10–21
Four-pole motor overload

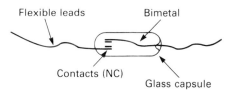

FIGURE 10–22
In-winding motor thermostat

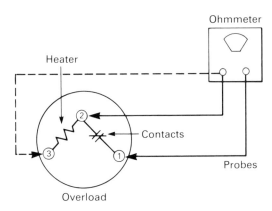

FIGURE 10–23
Checking external overloads with ohmmeter

To check an in-winding thermostat overload, open the disconnect to the motor power circuit, and disconnect the power leads to the motor. Make sure that the motor is cool before you make this check. Check for continuity between the motor's common (C) and run (R) terminals with an ohmmeter, as shown in Figure 10–24. If there is no continuity, proceed to check for continuity across the start (S) and common terminals. If there is no continuity at either of these points, either the overload is open or one or both of the motor windings have opens. Wait a sufficient length of time to be sure that the motor windings are not hot, and then recheck. If there is still no continuity, then check the motor windings (this procedure is discussed in Chapter 12). If the overload is open or the windings are burned out, the motor must be replaced.

The procedure for checking overloads in hermetically sealed compressor motors or any completely enclosed motor is covered in Chapter 12.

10.7

Summary

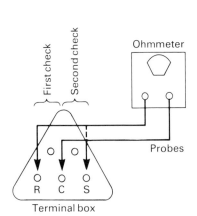

FIGURE 10–24
Checking internal overloads with ohmmeter

The proper operation of an air conditioning or heat pump system depends on the control system. Regulation of the cooling capacity to match the load and protection of the components against damage due to malfunction are both dependent on the controls.

The controls used in a system can be divided into two main categories: operating controls and safety controls. Operating controls include the transformer, thermostats, control relays, and liquid line solenoid valves. Safety controls include pressure switches and motor overloads. Each type of control has a special procedure for servicing and troubleshooting, and most of these procedures were described in this chapter.

In some cases, the description of the controls was included in this chapter, but their servicing techniques are covered more fully in other chapters in the text. This is the case when the operation of the controls is directly related to other main parts of the system. Consult those chapters, as referenced in the text, for a complete description of service procedures.

10.8

Questions

1. An air conditioning system is turned off and on by the control system to provide the _____ amount of cooling over a given period of time to _____ the cooling load of the building.
2. What are the two main categories of controls used in an air conditioning system?
3. Describe the function of the operating controls in a system.
4. Describe the function of the safety controls in a system.
5. A typical control transformer is a step- _____ transformer.
6. Is the following statement true or false? The primary side of a transformer is connected to the power source.
7. Is the following statement true or false? If the primary voltage on a transformer varies, the secondary voltage will vary proportionately.
8. A cooling thermostat has a switch that _____(opens, closes) on a temperature rise.
9. Describe the bimetal element used in a thermostat.

10. Describe the mercury bulb switch used in a thermostat.
11. Describe the anticipator used in a cooling thermostat.
12. A liquid line solenoid valve is used to control the flow of _____ in an air conditioning system.
13. Describe a pump-down control system.
14. Coils in controls can be checked for continuity by the use of an _____.
15. Contacts in controls can be checked with an _____ or with a _____.
16. A pressure switch is a control that senses _____ in some part of the system and operates an _____ switch in response to that pressure.
17. What is a motor overload used for in an electric circuit?
18. Name the three types of overload controls, as defined by the number of electric poles.
19. What is the function of an in-winding thermostat?

Electric Motors

11.1

Introduction

Electric motors are the most common source of operating power for air conditioning or heat pump systems. They are used to drive the compressor, condenser, and evaporator blowers. They are also used to operate dampers, valves, and other motorized parts of some systems. Thus, a service technician must understand how electric motors work and how they are built in order to service them properly.

This chapter gives a general description of the parts of a typical motor, describes the operating principles, discusses the types of electric motors used in air conditioning or heat pump systems, and discusses the controls that are an inherent part of the motors.

11.2

Description and Parts

An **electric motor** is a mechanical device that converts electric energy into mechanical motion. In an electric motor, electricity is used to produce magnetism. The magnetism is used to turn a shaft, and the turning shaft is used to do work.

Electric motors are the most popular and common type of motor used for providing mechanical power for air conditioning or heat pump systems. The main reasons for their popularity are the ready availability of

Electric motor

Frame

Stator

Field coils
Rotor

Armature

electric power and the simplicity of the motors. A typical electric motor is shown in Figure 11–1. Electric motors are relatively inexpensive and will operate continuously with a minimum amount of maintenance or service.

An electric motor is made up of three major parts and several minor parts. The three major parts are the frame, the stator (or field coils), and the rotor (or armature). Figure 11–2 is a cutaway view of an electric motor showing the major parts. The **frame** includes the outside enclosure, the end bells, and the parts that hold everything together. The **stator** is the part of the motor that fits inside the frame and surrounds the rotor. The major components of the stator are the field coils. The **field coils** generate the magnetic field that makes the motor run. The **rotor** is the part of the motor that turns. The major component of the rotor is the armature. The **armature** is the part of the rotor that reacts with the magnetic field generated in the field coils to produce rotation. The armature is mounted on the shaft and causes it to turn. As the rotor turns, the shaft turns also.

FIGURE 11–1
Typical electric motor for air conditioning or heat pump system (Courtesy of Franklin Electric Co., Inc.)

Stator

Field coils

Rotor

End bell

Armature

Frame

FIGURE 11–2
Cutaway view of electric motor (Courtesy of Franklin Electric
Co., Inc.)

The shaft runs in bearings that are mounted on the end
bells of the frame.

In addition to the main parts, most electric motors
used in air conditioning or heat pump systems have
controls that are inherent to their operation. These controls include starting switches or relays and start and
run capacitors. These controls are described in this chapter along with the motors. The service procedures related to both the motors and the controls are covered in
Chapter 12.

11.3

Electric Motor Operation

The field coils in an electric motor are coils of wire wound
around an iron form. The form is part of the stator, and
it forms a cylinder that the rotor turns in. The coils produce a magnetic field when an electric current flows

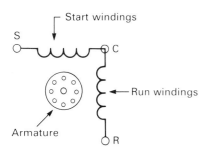

FIGURE 11–3
Schematic of electric motor

Induction motors

through them. The field coils are always arranged in pairs. Figure 11–3 is a schematic of the field coils and the armature of an electric motor of the type used in air conditioning and heat pump systems. In the figure, S represents the terminal for the start windings, C represents the common terminal, and R represents the terminal for the run windings. Start and run windings are described later in this chapter.

As electric current flows through the field coils, the magnetic polarity in the coils directly opposite each other is always exactly opposite. That is, if the pole on one side has north polarity, the pole opposite it has south polarity. As alternating current flows through the coils, the polarity changes with the alternations, and it changes in such a way that a rotating magnetic field is produced. As the magnetic field in the field coils rotates, it causes the armature—and, consequently, the motor shaft—to rotate also.

Most motors used in air conditioning or heat pump systems are **induction motors**. That is, the magnetic field in the field coils induces current flow in the coils that are included in the armature. The currents in the armature then produce magnetic fields of their own. The magnetic fields in the armature have polarity, just as the fields in the stator do. At any given time, the polarities of the magnetic fields in the stator and the rotor are aligned in such a way that the rotor turns in an attempt to bring the poles together. Figure 11–4 illustrates this relationship of the poles in the stator and the rotor. As the rotor turns and the magnetic poles start to come into alignment, the alternating current changes the polarity so that the armature continues turning. The armature and rotor turn as long as electric current is flowing in the coils.

11.4

Electric Motor Characteristics

Electric motors are defined by type according to specific characteristics. One way in which they are typified is by the electric current the motor uses. Another way is by the electric voltage, phase, and hertz (or cycles). A third

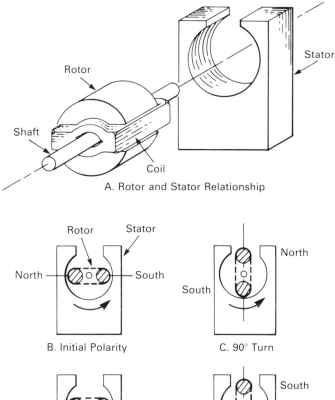

A. Rotor and Stator Relationship

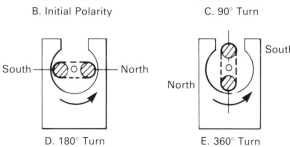

B. Initial Polarity

C. 90° Turn

D. 180° Turn

E. 360° Turn

FIGURE 11–4
Polarity in armature and field coils of motor

characteristic is related to the starting devices used on the motor. A fourth characteristic is the way the motor is connected to the load. Each of these characteristics is covered in the paragraphs that follow.

There are two types of electric current in general use: direct current (DC) and alternating current (AC). **Direct current** is the type of electricity that is available

Direct current

Alternating current

Cycles

Voltage

Phase

Single-phase power
Three-phase power

from batteries. It is called direct current because the electrons always flow in one direction in a circuit. Direct current electricity is seldom used in motors in air conditioning or heat pump equipment, so DC motors will not be covered in this text.

Alternating current is the type of electricity obtained from the typical electric supply system in most buildings. In an alternating current circuit, the electrons reverse their direction of flow a given number of times per second. The reversals are called **cycles**, and in the United States, this reversal occurs 60 times per second. The motors described in this text are all alternating current motors.

Voltage is a measurement of electric potential in a circuit. It can be likened to pressure. It is the force that moves electrons through a resistance, such as a wire. Voltage provides the electric power needed to operate a motor. Motors are available to operate on all of the readily available voltages, such as 115 volts, 230 volts, 208–230 volts, and so on.

Phase is a term that refers to the number of pulses of electric power that occur in each cycle. The two common phases of power are single phase and three phase. **Single-phase power** has two pulses per cycle, one positive and one negative. **Three-phase power** has six pulses per cycle, three positive and three negative. Figure 11–5A shows a sine wave graph for a single-phase power source (E is the symbol for potential). Figure 11–5B shows the sine wave graphs for a three-phase power source.

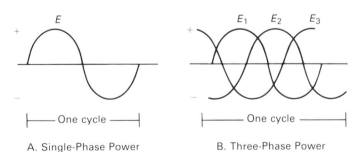

A. Single-Phase Power   B. Three-Phase Power

FIGURE 11–5
Sine wave graphs of electric power

Either single-phase or three-phase power may be used in an air conditioning or a heat pump system. Single-phase power is usually used for small units, and three-phase power is used for large units.

Hertz (Hz) is a term related to the rate of alternations in the voltage and current in the electric supply system. In the past, the term used for this rate was cycles. The alternating current produced in the United States cycles—or goes from positive to negative and back to positive on a sine wave graph—at a rate of 60 times per second. In countries using the metric system, it cycles at a rate of 50 times per second. Power in the United States is 60-hertz power, and in most other countries, it is 50-hertz power.

The third characteristic by which motors are typified is the method used to start them. While single-phase motors do have an automatically rotating magnetic field, because of the way the coils are arranged, they do not start by themselves. A starting device of some sort has to be used to start rotation when the power is first turned on. Three-phase motors have three field coils in their stators, as shown in Figure 11–6. The coils are arranged so that the three phases of power in each cycle create a rotating magnetic field. This rotating magnetic field causes the rotor to turn naturally. The naturally rotating magnetic field in a three-phase motor makes a starting switch unnecessary.

The fourth method of typifying motors is by the way the motor is connected to the load it is driving. There are two basic types of drives used: direct drive and belt drive. When an electric motor is mounted externally to the equipment that it is driving, and the connection between the motor and the apparatus is by pulleys and belt, it is called a **belt-drive system**. This system is shown in Figure 11–7A. If the motor shaft is connected directly to the shaft of the driven machine, then the system is called a **direct-drive system**. Figure 11–7B shows the direct-drive system.

Belt-drive arrangements are used on driven devices when the rotational speed of the device may need to be adjusted. Changing the pulley sizes changes the speed of rotation of the driven device. If the speed of rotation of a device connected by a direct drive must be changed, then a multispeed motor is used. The speed of the motor

Hertz

Belt-drive system

Direct-drive system

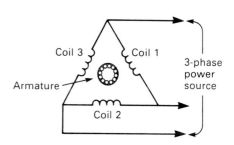

FIGURE 11–6
Schematic of three-phase motor

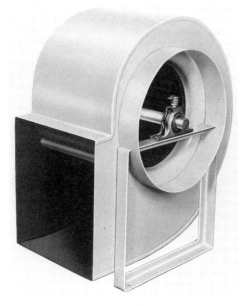

A. Belt drive

B. Direct drive

FIGURE 11–7   Types of drives for motors (Courtesy of Barry Blower Co.)

is controlled by the way it is connected to the electric power supply.

In air conditioning or heat pump systems, direct-drive motor connections are often used in hermetic or semihermetic compressors. In this application, the compressor and the motor are connected on a common shaft, and both are placed in a hermetically sealed container. The motor is specially built so that the motor frame is part of the compressor frame. Figure 11–8 is a cutaway view of a hermetic compressor, showing the motor inside the container.

11.5

Motors Used in Air Conditioning Systems

Electric motors are built in different sizes according to the amount of mechanical power they will produce. They are rated in horsepower (hp). The term **horsepower** is

Horsepower

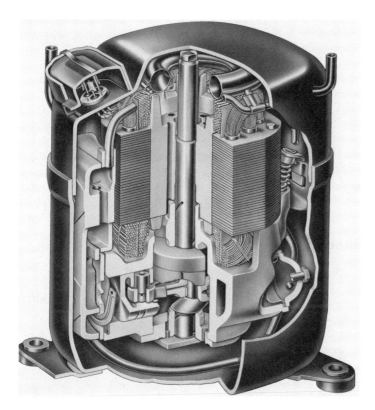

FIGURE 11–8
Cutaway view of hermetic compressor (Courtesy of Tecumseh
Products Co.)

related to mechanical work, and one horsepower is the
equivalent of 33,000 foot pounds (ft-lb) of work. The
rating is used for electric motors because electric energy
is changed to mechanical work in a motor.

The method used for starting an electric motor determines the mechanical power on start-up, in compar-
ison with the running power. Motor selection for an air
conditioning or a heat pump system is made according
to both the starting power and the running power required.

Six basic types of motors are used in air conditioning or heat pump systems, as categorized by starting
types. The first five motors are single-phase motors, and
the sixth motor is a three-phase motor. Each single-phase

motor has a different type of starting device. The six types of electric motors are as follows:

1. Shaded pole motor
2. Split-phase motor
3. Permanent split-capacitor (PSC) motor
4. Capacitor start (CS) motor
5. Capacitor start, capacitor run (CSCR) motor
6. Three-phase motor

Each type of motor is described in the sections that follow.

Shaded Pole Motor

Shaded pole motor

Shading ring

A **shaded pole motor** is the simplest of the induction motors. A shaded pole motor has a single pair of field coils. Each coil face has a copper ring, called a **shading ring**, around part of it, as illustrated in Figure 11–9. With this feature, the motor does not need a starting switch or relay.

Figure 11–10 is a schematic of a shaded pole motor. When an electric current flows through the field coil, the face of the coil is magnetized. An electric current is induced in the shading ring by the magnetic field. But

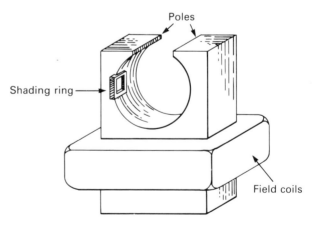

FIGURE 11–9
Shading ring on field coil

the current in the shading ring flows in the opposite direction to the current in the coil. This flow is in accordance with a law of electric induction. The current in the shading ring causes the magnetic field within the ring to be out of phase with the magnetic field in the rest of the coil face. The out-of-phase condition creates a rotational effect on the rotor of the motor. This rotational effect starts the motor when the electricity is turned on, and it has no noticeable effect while the motor is running.

Shaded pole motors are usually used only in the small horsepower sizes, motors of 1/4 horsepower and less being the most common. The power developed on start-up with a shaded pole motor is about the same as the running power. Shaded pole motors are used when the motor does not have to start against a load or run under a very heavy load. They are commonly used for propellor types of blowers, where the resistance to air movement is small.

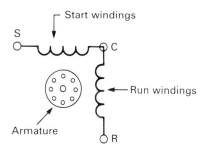

FIGURE 11–10
Schematic of shaded pole motor

Split-Phase Motors

Split-phase motors are used on many air conditioning or heat pump applications. A split-phase motor has two sets of field coils. One set of coils is used for running the motor. These coils are called **run windings**. The other set of coils is used for starting the motor. These coils are called **start windings**. The start windings are physically located in the stator so that they are in between the run windings, as shown in Figure 11–11.

The coil wire used in the start windings is of a much smaller diameter than the coil used in the run windings, and consequently, there is more of it. The smaller size and extra length of the wire in the start windings causes more resistance to the flow of electric current. The extra resistance causes an electrical phase difference between the start windings and the run windings. This phase difference causes the motor to turn and also increases the torque on start-up. The starting torque of a split-phase motor is approximately 150% of the running torque.

On a split-phase motor, if the start winding is allowed to remain energized after the motor comes up to

Split-phase motors

Run windings

Start windings

FIGURE 11–11
Start and run windings of split-phase motor

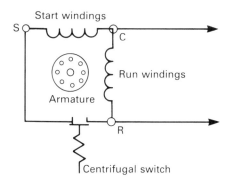

FIGURE 11–12
Schematic of split-phase motor

operating speed, the coil overheats because of the high resistance. In this situation, the coil will burn out. A starting switch or relay is used to take the start winding out of the circuit when the motor has nearly reached its normal operating speed. The starting switch may be a centrifugal switch, or it may be one of several kinds of starting relays. The starting switch opens a set of contacts that are wired into the start winding circuit when the rotational speed of the motor shaft reaches approximately 75% of its normal running speed. Figure 11–12 is a schematic diagram of a split-phase motor with a starting switch wired into the start winding circuit.

Split-phase motors are used on applications that do not require a great deal of starting torque. They are often used on evaporator or condenser blower motors, because these motors do not have to start against a heavy load. They are also used on small compressors for air conditioning or heat pump systems with expansion devices that do not shut off during an off cycle of the unit. The starting load is small in this application.

Permanent Split-Capacitor Motor

Permanent split-capacitor motor

A **permanent split-capacitor motor** is similar in construction to a split-phase motor. It has two sets of coils that are physically and electrically out of phase with each other. This out-of-phase condition provides starting rotation and increases starting torque. The starting torque is further increased by a capacitor wired in series with the start winding. This connection places the capacitor in parallel with the run winding, as shown in Figure 11–13. The capacitor provides additional starting torque and also improves the efficiency of the motor by improving the power factor during normal running time. The starting torque of a permanent split-capacitor motor is approximately 150% of the running torque.

As the name of the motor implies, the capacitor in a permanent split-capacitor motor stays in the start circuit continuously. A starting switch is not used. Thus, the motor must use a type of capacitor that will work continuously without burning out. This type of capacitor is called a run capacitor. A run capacitor of the proper rating can remain in the circuit during the operation of the motor without causing damage to the capacitor or

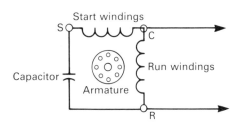

FIGURE 11–13
Schematic of permanent split-capacitor motor

the motor. Permanent split-capacitor motors have a somewhat higher starting torque than split-phase motors. But they are still used on blowers and small compressors that do not have to start against heavy loads.

Capacitor start motor

Capacitor start, capacitor run motor

Three-phase motors

Capacitor Start Motor

A **capacitor start motor** is similar in operation to a permanent split-capacitor motor. However, a starting switch is installed in the start winding circuit along with the capacitor, as shown in Figure 11–14. Also, a start capacitor is used instead of a run capacitor. These features give the motor more torque on start-up than a split-capacitor motor has. The starting switch takes the capacitor and the start winding out of the circuit when the motor comes up to about 75% of its running speed. The starting switch can be either a centrifugal switch or one of the different types of starting relays.

A capacitor start motor develops approximately 200% of its running torque on start-up. Capacitor start motors are normally used in applications where the motor has to start in a partly loaded condition. Some blower motors on large air movers and the compressors on systems with expansion valves use capacitor start motors.

Capacitor Start, Capacitor Run Motors

A **capacitor start, capacitor run motor** is one that has one or more capacitors in both the run and the start circuits. In this type of motor, start capacitors are used in the start circuit, and run capacitors are used in the run circuit. The start capacitor (or capacitors) is wired in series with a starting switch so that it can be taken out of the circuit when the motor comes up to speed. The run capacitor remains in the circuit permanently. The schematic for this motor is presented in Figure 11–15.

Three-Phase Motors

Three-phase motors are simpler in construction than single-phase motors. Three-phase electric power is transmitted through three separate wires, with each wire

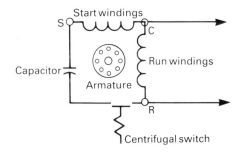

FIGURE 11–14
Schematic of capacitor start motor

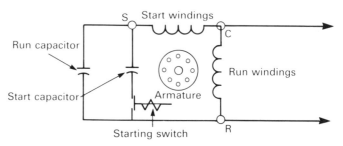

FIGURE 11–15
Schematic of capacitor start, capacitor run motor

carrying one phase. The motor is constructed with three field coils (see Figure 11–6). It is wired in such a way that each coil is energized by a different wire in the three-phase circuit. Each coil is actually 120 degrees out of phase with the coil preceding it, both physically and electrically. As the current flows through each coil in turn, a rotating magnetic field results.

Three-phase motors are normally built in large sizes. They are readily available in sizes from 1 horsepower and higher. Most compressors requiring 3-horsepower or larger motors are three-phase. The starting torque on a three-phase motor is significantly greater than the running torque. Also, starting switches do not have to be used on three-phase motors. Finally, motor rotation can

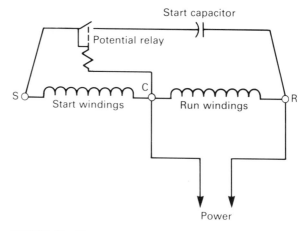

FIGURE 11–16
Potential relay used as starting switch

be reversed simply by changing the positions of any two of the three power leads to the motor.

11.6

Starting Switches

Many electric motors used in air conditioning or heat pump systems require **starting switches**. These switches are used to remove the start windings and the capacitors from the start circuit after the motor reaches about 75% of its rated speed. The most commonly used starting switches are centrifugal switches and starting relays. Starting relays include potential relays, current relays, hot-wire relays, and solid-state relays. These devices are described in the following paragraphs.

Starting switches

A **centrifugal switch** is a device mounted on the shaft of the motor inside one of the bell ends. As the shaft rotates, two counterbalanced weights are forced away from the shaft by centrifugal force. The weights are attached to a sliding ring by connecting links. As the weights move, the ring opens an electric switch. The switch is wired in series in the starting circuit. The switch is adjusted so that it opens when the motor is running at about 75% of its rated speed.

Centrifugal switch

Starting relays are similar to regular relays in that they have a coil and a set of contacts that open or close as the coil is energized. A **potential relay** is energized by voltage. When a potential relay is used as a starting relay, it is wired into the motor circuit so that the coil senses the voltage, or potential difference, across the starting coil. Figure 11–16 shows the wiring connections for a potential relay used as a starting relay. When the motor comes up to about 75% of its rated speed, the voltage across the starting coil is great enough to cause the contacts in the relay to open. When the contacts open, the start winding and the capacitor, if there is one, are taken out of the starting circuit.

Starting relays

Potential relay

A **current relay** is energized by amperage. As shown in Figure 11–17, a current relay is wired into the motor circuits with the coil in the power leg to the run terminal. The contacts are wired in the start circuit. On motor start-up, the relay is closed. But as the motor comes up

Current relay

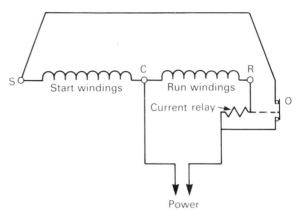

FIGURE 11–17
Current relay used as starting switch

to speed, the current flowing through the relay coil increases until the contacts open. When the contacts open, the start winding is taken out of the circuit.

Hot-wire relay

A **hot-wire relay** operates very much the same way as a typical time delay relay. It has a small electric heater enclosed with a bimetal switch. As Figure 11–18 shows, the heater in the relay is wired in series with the hot leg to the run terminal on the motor. The bimetal switch is wired in series in the start circuit. The contacts are normally closed. When the motor starts, current flow through the heater causes it to heat up. This heating

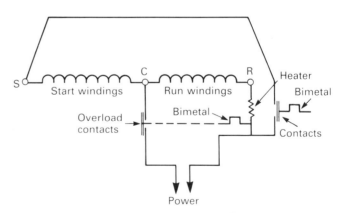

FIGURE 11–18
Hot-wire relay used as starting switch

opens the bimetal switch. The relay is designed so that the switch will open when the motor has come up to about 75% of normal speed. Most hot-wire relays have a second set of contacts that act as a current overload. This set of contacts is in series with the power lead from the relay to the common terminal on the motor. If the current exceeds a set limit, this set of contacts opens and the motor stops.

Solid-state starting relays are becoming popular for many small-motor applications. A solid-state relay uses a self-regulating, conducting ceramic element in which the resistance changes as the current flow changes. This element acts as a switch that opens the start circuit on start-up. A solid-state starting switch and a capacitor can be added to some split-phase motors to change them to capacitor start motors.

Solid-state starting relays

11.7

Capacitors

A **capacitor** is an electric control device that consists of two conductor plates that are in close proximity but are insulated from each other. Figure 11–19 shows a schematic representation of a capacitor. Each plate is connected to one of the leads in an AC circuit. The flow of electrons in the circuit creates a potential difference between the two sets of plates. The potential difference on the plates is maintained because they are close together, but current cannot flow between the plates because of the dielectric insulation. In this condition, the capacitor is said to be charged.

As the current alternates—and capacitors are only effective in AC circuits—the polarity on the plates is reversed during each alternation. The alternate charging of the plates allows electric potential to be stored momentarily on each plate. But as the reversal of the current occurs, the stored energy is added to the energy being supplied by the current. So, additional voltage is available momentarily in the circuit.

There are two basic types of capacitors used in air conditioning and heat pump systems: run capacitors and start capacitors. They may be used in either or both run and start circuits of some motors. Each application re-

Capacitor

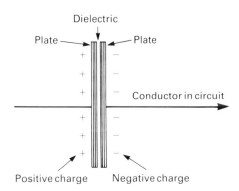

FIGURE 11–19
Schematic capacitor

Run capacitors

Start capacitors

quires a different type of capacitor. **Run capacitors**, shown in Figure 11–20A, are used primarily to improve the power factor of the motor during normal running. Thus, they make the motor more energy-efficient. **Start capacitors**, shown in Figure 11–20B, are used to increase the starting torque of the motor on start-up. In some cases, only one type of capacitor is used on a motor. In others, both are used.

As illustrated in Figure 11–21, run capacitors are wired in parallel with the run windings in a motor, and they stay in operation while the motor is running. Start capacitors are wired in parallel with the run winding but in series with the start winding. Hence, they are always taken out of the circuit when the motor comes up to normal speed.

11.8

Summary

Electric motors operate on an electromagnetic principle. Electricity is used to produce a magnetic field. This field

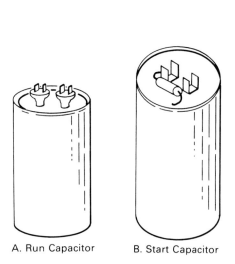

A. Run Capacitor   B. Start Capacitor

FIGURE 11–20
Run and start capacitors

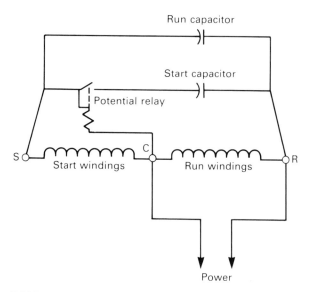

FIGURE 11–21
Wiring diagram for run and start capacitors in motor circuit

is then used to rotate an armature. The armature is attached to a shaft. As the armature rotates, so does the shaft. The shaft can be connected to any piece of machinery that works with a rotary motion. The motor provides power for the operation of the machinery.

Motors consist of three major parts and several minor parts. The major parts are the frame, the stator, and the rotor. Some of the minor parts are the bearings, the shaft, and starting switches and relays.

The major types of motors used in air conditioning or heat pump systems are shaded pole motors, split-phase motors, permanent split-capacitor motors, capacitor start motors, capacitor start, capacitor run motors, and three-phase motors. Each of these types has different power characteristics, and most of them are controlled by different methods.

Most single-phase motors require starting switches to start them rotating. The switches keep the start windings—and in some cases, the capacitors—in the start circuit during start-up. But the switches remove the windings from the circuit after the motor reaches its operating speed. Centrifugal switches and various types of starting relays are used for this purpose. The most commonly used starting relays are potential relays, current relays, hot-wire relays, and solid-state relays.

Capacitors are used both for improving running characteristics of motors and for producing more starting torque. Two types are available: run capacitors and start capacitors. Each must be used for its own particular purpose.

11.9

Questions

1. An electric motor is a mechanical device that converts _____ energy to _____ motion.
2. Name the three major parts of an electric motor.
3. Which of the three main parts of an electric motor moves in a rotary motion?
4. Name the special device needed to start most single-phase motors.

5. A magnetic field is produced when a motor is turned on. What is the part called that produces the field?
6. What type of motor is used mostly in air conditioning systems?
7. List the four characteristics used to specify electric motors.
8. Name the six common types of motors used in air conditioning systems, as categorized by starting device.
9. Which of the six common types of motors develops the most power on start-up? Which develops the least?
10. Explain why a three-phase motor does not need a starting switch.
11. Name four types of starting relays or switches used on induction motors.
12. Capacitors are used on electric motors for two purposes. Name the types of capacitors used and the purpose of each.

Servicing Electric Motors and Controls

12.1

Introduction

Electric motors are used for at least three purposes in a typical air conditioning or heat pump system. They are used as condenser blower motors, as evaporator blower motors, and as compressor motors. The condenser and evaporator blower motors may be either belt-drive or direct-drive motors. The compressor motor may be either a belt-drive or a direct-drive motor, or it may be part of the compressor itself. In a hermetic or semihermetic compressor unit, the motor is built into the compressor frame and is enclosed in the housing with the compressor. The operation of the motor is similar in all of these cases, but the type of motor and the method of controlling it may vary.

Specific steps to take in servicing the various types of motors and motor controls are covered in this chapter. The chapter begins with a description of service of the mechanical parts of a motor. Then, service of the electric parts, including windings, overloads, start devices, and capacitors, is covered. Finally, service of hermetic compressor motors is discussed.

12.2

Mechanical Parts of a Motor

Motors can suffer from either mechanical failure or electrical failure of the components. When an electric motor is suspected of being the cause of an air conditioning or

Mechanical failure

External physical damage

Internal physical damage

a heat pump system's failure, the motor should be checked for mechanical faults first. **Mechanical failure** is evidenced by physical damage to the motor parts or by worn-out bearings or parts, and it can be either external or internal.

External physical damage is usually caused by fire, water, or some other factor outside the motor itself, and it is usually quite obvious. A visual inspection will normally identify any external physical damage to a motor. If minor cleaning and adjusting will make it run again, this service can be taken care of in the field. The motor can then be placed back into operation. If a motor is damaged enough to require dismantling for repairs, it is usually better to replace it than to try to repair it in the field.

Internal physical damage is caused by dirt, dust, water, or other contamination getting inside the motor. Burned-out bearings can result from dirt, sand, or other foreign material in the bearings or from lack of lubrication. To check for internal physical damage, remove the motor's housing and make a visual inspection of the parts of the motor. Figure 12–1 illustrates some of the

FIGURE 12–1
Cutaway view of electric motor showing parts (Courtesy of Franklin Electric Co.)

parts that can be checked. If a motor requires disassembling for repairs, it should be taken to a shop rather than being repaired in the field. If some part of the motor is found to be faulty, the part should be replaced.

Worn-out bearings are the most common cause of internal physical failure in electric motors. To check a motor for bad bearings, disconnect the motor from the electric power and turn the motor shaft by hand. If the shaft is difficult to turn, the bearings are probably worn or seized-up. If the shaft does turn easily, the bearings should be checked for end play by moving the shaft back and forth in a direction parallel with the axis of the shaft. Excessive end play indicates that the thrust bearing in the end of a single-end motor, or both bearings in a double-end motor, may be worn out. If any of the bearings appear to be worn out, they should be replaced before the motor is put back in service.

Worn-out bearings

12.3

Electric Power Supply

If an electric motor is found to be all right physically, but it still does not run, the trouble may be with some of the electric components of the motor. These components include the controls or devices that are directly connected with the operation of the motor. The electric components are the electric power supply to the motor, the overloads, the motor windings, the starting devices, and the capacitors. The electric power supply is discussed in this section. The other components, in the order listed, are covered in the following sections.

If you suspect that there may be some problems with the electric parts of a motor, the first check to make is to see whether there is electric power to the motor. To make this check, close the unit disconnect. Check the system controls that call for motor operation to make sure that the motor should be running. Then, use a voltmeter to check the voltage at the motor terminals. The two probes of the meter are placed on the electric terminals where the power circuit is connected to the motor, as illustrated in Figure 12–2. The voltage reading on the meter should be within ±10% of the rated voltage of the system.

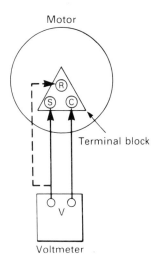

FIGURE 12–2
Checking voltage at motor with voltmeter

If the voltage at the motor terminals is the correct voltage for the system, but the motor still does not run, the motor overloads should be checked.

12.4
Overload Controls

After the power supply to the motor has been verified, the next component to check is the overload on the motor, if it has one. Motor overloads were described in Chapter 10. A general description is also given here, but refer to Chapter 10 for the methods to use in servicing them.

Overload protectors

Overload protectors are heat-sensing devices that open an electric circuit if the temperature around them exceeds the temperature for which they are calibrated. As shown in Figure 12–3, the overload has a bimetal element that warps when it is heated. The bimetal element has a normally closed (NC) contact that is mated to another contact. The contacts are wired into either the power circuit for the motor or the motor control circuit. If the overload senses a high temperature, the contacts open as the bimetal warps, and the circuit is broken. The broken circuit shuts the motor off.

Some overload protectors reset themselves when the motor temperature returns to normal, but others are built so that they must be reset manually. If a motor cycles off and on, and especially if it seems to be running hot, the amperage draw of the motor should be checked. If the motor is overloaded, it will draw more amperes than it is rated for, and the overloads will open. If the motor goes off unexpectedly and will not come back on, the overloads should be checked. Remember that if a motor is cycling on the overloads, this operation does not indicate that something is wrong with the motor. Rather, it indicates that something is wrong with the loading of the motor or with the motor circuits.

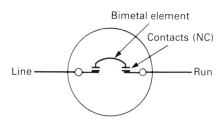

FIGURE 12–3
Two-wire overload for electric motor

12.5
Motor Windings

If the overload control devices in the motor are not at fault, the next series of checks should be made on the motor windings. Different manufacturers use different numbers or letters for identification of the **coil terminal connections** on their motors. The most commonly used identifications for the terminals on a single-phase motor are C for common, S for start windings, and R for run windings, as indicated in Figure 12–4. Three-phase motor terminals normally are identified as 1, 2, and 3. Refer to the wiring diagram inside the cover plate on the motor terminal box to determine what identification is used in the motor you are working on. The terminal box is usually located on one of the end bells of the motor. In this text, we will use the terminal designations given above.

Most motor winding failures occur because of overheating. Excessive heat breaks down the insulation on the coil wires. Then, either the coil shorts out internally, or a short occurs between the coil and the motor frame. If the short is internal, it shows up as an open in the circuit or as an increase in the resistance through the windings. In checking the motor windings, you must check for each type of short. The procedures for these service checks for both single-phase and three-phase motors are described in the following subsections.

Checking Single-Phase Motor Windings

To check a single-phase motor for an open in the windings, first make sure that the power is turned off at the unit disconnect. Then, disconnect the power leads from the motor terminals. Identify which terminals are related to the start windings and which to the run windings. If the terminals cannot be matched to the windings from inspection of the terminal block, then the motor may have to be dismantled and the leads from the coils located inside.

Using an ohmmeter, check the start windings for continuity. As illustrated in Figure 12–5, place one of the probes of the ohmmeter on the C (common) terminal, and place the other probe on the S (start) terminal. A reading of infinite resistance, which indicates no continuity, indicates an open in the coil windings. A reading of a fairly low but still measurable resistance indicates that the motor windings are good.

Follow the same procedure for the run windings. Place one probe of the meter on the C terminal and the

Coil terminal connections

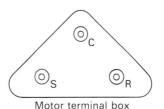

Motor terminal box

FIGURE 12–4
Typical motor terminal box

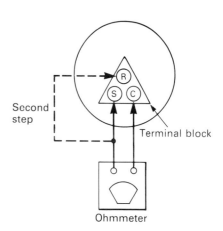

FIGURE 12–5
Checking motor windings for opens

other on the R terminal. The meter readings indicate the same things for this coil as they do for the start coil. An open in either case indicates that the motor should be rewound or replaced.

Electric motor windings should also be checked for the resistance, in ohms, through each winding. The resistance value for proper operation should be stated by the manufacturer of the motor. To check the resistance, make sure that the power is turned off at the unit disconnect switch. Then, disconnect the power leads from the motor terminals. Use an ohmmeter that is sensitive enough to give a reading within 5 ohms, and read the resistance between the R and the C terminals. Then, read the resistance between the C and the S terminals. (See Figure 12–5 for the procedure.) The resistance reading between the C and the S terminals should be greater than the reading between the R and the C terminals.

The next step is to read the resistance between the R and the S terminals. This reading should equal the sum of the readings taken between R and C and between C and S.

These two relationships are shown in the following formulas.

resistance between R and C < resistance between C and S

resistance between R and S = (resistance between R and C) + (resistance between C and S)

These readings should also be within the allowances given by the manufacturer for the resistance through each of the coils. If they are not, then the motor should be rewound or replaced.

To check for an internal short to ground, one probe of an ohmmeter is placed on one of the motor terminals. The other probe is placed on the shell or frame of the motor, as indicated in Figure 12–6. If the ohmmeter reads less than 3 million ohms, then a short to ground is indicated. The motor needs rewinding or replacing. Use the same check for each of the windings.

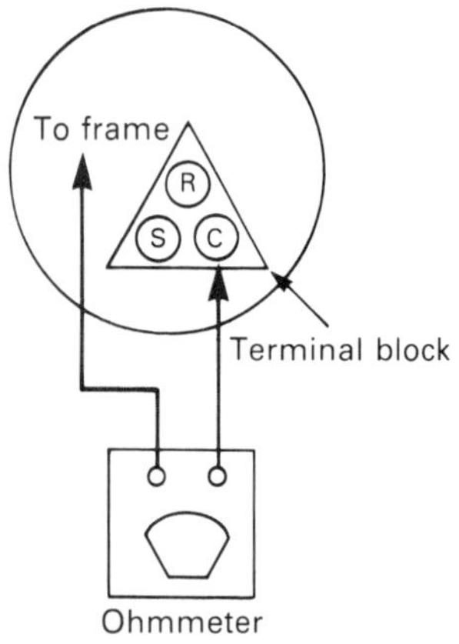

FIGURE 12–6
Checking motor coils for internal short to ground

Note: These checks should only be made with a cool motor. If there is an internal overload, and if the motor is hot, the overload may be open.

Checking Three-Phase Motor Windings

To check a three-phase motor for internal opens, open the unit disconnect. Then, disconnect the main power leads from the motor terminals. With an ohmmeter set on the ×10K scale, check all three combinations of terminals 1, 2, and 3. Check 1 to 2, 1 to 3, and 2 to 3. If there is an open, indicated by no continuity between any pair of terminals, the motor will have to be rewound or replaced.

To check the resistance through the motor windings, use the same procedure used for checking the windings of a single-phase motor, but check between each set of terminals. If the readings are not within the resistances specified by the manufacturer of the motor, then the motor will have to be rewound or replaced.

To check a three-phase motor for a short to ground, place one probe of the ohmmeter on the frame of the motor and the other probe on each terminal, in turn. If there is a reading below 3 million ohms between any terminal and the frame, there is a short. The motor will have to be rewound or replaced.

12.6

Starting Switches

With one exception, the single-phase induction motors used in air conditioning or heat pump systems have a set of start windings. The exception is the shaded pole motor. The start windings in these motors provide rotational torque to the rotor on start-up. But in most of these motors, the start winding must be taken out of the circuit when the motor comes up to its operating speed. A device called a starting switch is used to take the start windings out of the electric circuit when the motor is nearly up to normal running speed.

Different types of devices are used as starting switches on different motors. Among these devices are centrifugal switches, potential relays, current relays, and solid-state relays. The servicing procedures for each device are described in the following subsections.

Centrifugal Switches

A centrifugal switch is shown in Figure 12–7. When a centrifugal switch fails, the contacts either will not close when the motor is energized or will fail to open when the motor comes up to operating speed. Failure of the contacts to close is usually due to mechanical problems within the switch. The contacts fail to open normally because they have become welded shut from excessive current flow.

The best way to check the switch for mechanical failure is by visual inspection. You may be able to see the switch through the ventilation slots in the end bell of the motor. But you will probably have to take the

Centrifugal switch

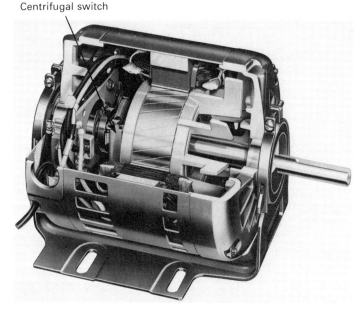

FIGURE 12–7
Centrifugal switch used in electric motor as starting switch
(Photo courtesy of Marathon Electric Mfg. Corp.)

motor apart to determine whether the switch is operating properly. Dismantling the motor should not be attempted in the field; it should be done only in a shop. With the motor apart, the operation of the switch can be observed.

To take a motor apart, turn off the power to the unit, and disconnect the wiring and control connections. The motor should be removed from the unit it is attached to. The end bell should be removed from the end of the motor in which the switch is located. The centrifugal switch is attached to the inside of the end bell.

With the motor apart, you should be able to see the working parts of the switch. A pair of weights on a ring slides on the motor shaft, and the switch is opened and closed by the motion of the ring. If the contacts are not closed when the weights are in the position they would be in when the motor shaft is not turning, something is wrong with the switch. If there is some mechanical breakage, it should be apparent.

The contacts should be closed with the centrifugal weights in the normal position. Move the weights away from the position they would normally be in when the motor is not running, and observe whether the contacts open. If the contacts do not open when the centrifugal weights are moved away from the shaft, the contacts are probably welded shut. If the switch is not closed when the weights are at rest, and is open when the weights are moved away from the shaft, then the switch is faulty and should be replaced before the motor is used again. A faulty starting switch will usually cause the start windings in a motor to short out.

Potential Relays

A potential relay has a set of contacts and a coil similar to those elements in a magnetic relay. Figure 12–8 is a schematic of a potential relay. One of the terminals in a potential relay is common to both the coil and the contacts. The coil in a potential relay senses voltage, and the contacts open when the voltage through the coil reaches a predetermined magnitude.

In use, a potential relay is wired into the motor circuit so that the coil is wired in parallel with the start

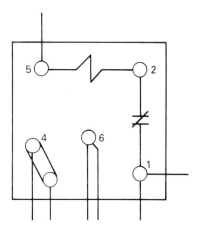

FIGURE 12–8
Schematic of potential relay

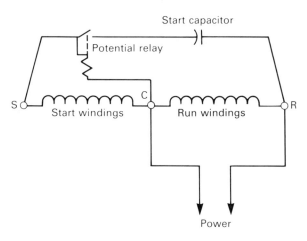

FIGURE 12–9
Schematic of potential relay in motor circuit

windings and so that the contacts are between the S and the R terminals. This wiring arrangement is shown in Figure 12–9. The coil is energized by the voltage generated by the start windings when the motor has come up to about 75% of its rated speed. The contacts open when this voltage reaches the predetermined voltage that the relay is designed for. The voltage in the coil circuit may be greater than the line voltage to the unit.

To check a potential relay coil, first disconnect the coil from the circuit. Then, using an ohmmeter, check for continuity across the coil. Put one of the probes of the meter on each terminal of the coil, as shown in Figure 12–10. If the meter shows that there is continuity, then the coil is good. If the meter indicates that there is an open (that is, no continuity), then the relay should be replaced.

To check the relay contacts, use a heavy-gauge, insulated wire jumper. Turn on the power to the unit, and set the thermostat to call for operation of the unit. When the motor attempts to start, it will hum but not start. Use the jumper across the contacts on the relay by placing the jumper on the terminal on one side of the contact and on the terminal on the other side, as shown in Figure 12–11. Hold the jumper in place until the motor starts or until it becomes obvious that the motor will not start. If the motor starts, remove the jumper. If the motor continues to run with the jumper

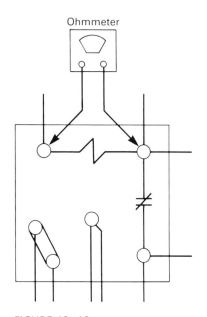

FIGURE 12–10
Checking potential relay with ohmmeter

off, then the potential relay is faulty and should be replaced.

Current-Sensing Relays

Often, a current-sensing relay is used as a starting switch. The current relay is wired into the starting circuit of the motor so that the coil of the relay is in series with the power lead to the run windings terminal, as shown in Figure 12–12. The contacts are wired so that they are in the circuit that goes from the same power lead to the start terminal of the motor.

    If the current-sensing relay is suspected of being faulty, it can be checked by the same procedures as are used for a potential relay. After turning off the power, remove the relay from the circuit by disconnecting the wires to it. Check for continuity across the heater with an ohmmeter. If the heater proves to be good, by showing continuity, reconnect the wires to the terminals. Now, check the operation of the relay when the motor is energized. In this check, use a heavy, insulated wire jumper across the terminals of the contacts. The contacts are normally closed on start-up, and the jumper wire will replace the closed contacts. If the motor starts with the jumper in place, remove the jumper. If the motor con-

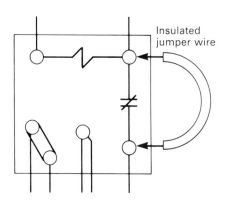

FIGURE 12–11
Using jumper wire to check potential relay contacts

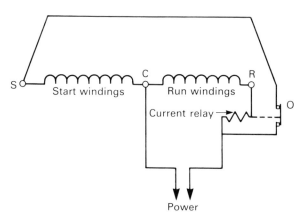

FIGURE 12–12
Current relay as starting switch

tinues to run, the current-sensing relay is bad and should be replaced.

One further check of the relay can be made. In this check, you observe the operation of the relay contacts with a voltmeter while the motor is turned on. One of the probes of the meter is placed on each terminal of the relay contacts, and the voltage is observed as the motor tries to start. When the contacts are closed, the meter shows no voltage. If the contacts open when the motor is nearly up to running speed, then the meter will show a voltage. A voltage reading indicates that the relay is functioning correctly.

Solid-State Relays

To troubleshoot a solid-state starting device, follow the instructions given by the manufacturer of the control. As with all other relays, if the relay is found faulty, you should replace it rather than try to repair it. Use a replacement with the same model number and rating as the relay removed from the unit.

12.7

Capacitors

Capacitors are often found in the motor circuits of air conditioning or heat pump systems. Two types are used: start capacitors and run capacitors. Start capacitors are used to provide increased starting torque. Run capacitors are used to improve the power factor of the motor. An improved power factor makes the motor operate more efficiently. Both capacitors are important to the operation of any system they are installed on. Figure 12–13 shows both capacitors used in an electric motor circuit.

To check for a faulty capacitor in a system, first make a visual inspection of the capacitor. Bulges in the capacitor container or leaks in the seams indicate a bad capacitor. If either fault is seen, the capacitor should be replaced with one of the same type and electrical rating.

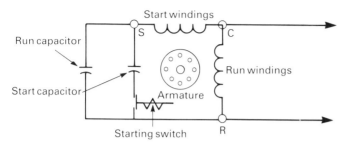

FIGURE 12–13
Start and run capacitors in electric motor

If no visual damage to the capacitor is apparent, then a **capacitor analyzer** can be used to check its internal condition. For this check, turn off the power to the unit. Short across the capacitor terminals with a heavy, insulated wire jumper in order to equalize the charge on the plates. Plug the analyzer into a power outlet. Place the probes of the analyzer on the terminals of the capacitor, as shown in Figure 12–14. If the capacitor is good, the indicator on the analyzer will show that condition. If the analyzer indicates that the capacitor is bad, it should be replaced.

Capacitor analyzer

If a capacitor analyzer is not available, an ohmmeter can be used to check a capacitor. An ohmmeter set on the ×10K scale is used for the checks. Disconnect the capacitor from the circuit, and short across the terminals to discharge the capacitor. Place the probes of the ohmmeter on the capacitor terminals, as illustrated in Figure 12–15. Observe the movement of the needle on the meter. On a good capacitor, the needle should swing to zero resistance and then slowly move partway back across the scale toward the infinity mark. On a shorted capacitor, the needle will swing to zero resistance and stay there. A capacitor with an open in it will show no movement of the meter needle. If the capacitor is shorted or has an open, it should be replaced.

Safety note: Start capacitors usually have a bleed resistor wired across the terminals in order to bleed off the electric charge during an off cycle. (See Figure 12–15.) This resistor should be disconnected from one end to perform the capacitor checks.

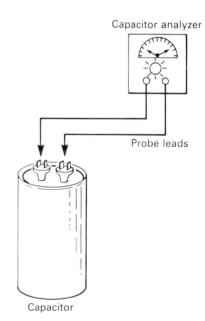

FIGURE 12–14
Checking capacitor with capacitor analyzer

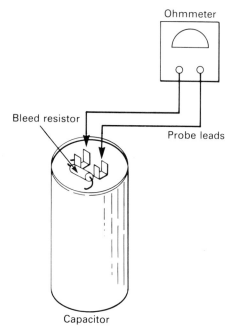

Ohmmeter

Bleed resistor

Probe leads

Capacitor

FIGURE 12–15
Checking capacitor with ohmmeter

12.8

Hermetic Compressor Motors

Electric motors used in hermetic or semihermetic compressors have the same parts and operating controls as other motors. The difference between hermetic compressor motors and other motors is that hermetic compressor motors are built into the compressor frame and are sealed inside the compressor shell. In a hermetic compressor, the motor is not accessible for service. And it is only accessible in a semihermetic compressor if the compressor is disassembled.

Centrifugal switch starters are not used on hermetic or semihermetic compressor motors. Some type of starting relay is used as a motor starting switch. All of the operating controls for hermetic compressor motors are on the outside of the compressor shell. The electric power leads to the motor pass through the hermetic shell through sealed terminals. The electric terminals for single-phase motors are identified by C, S, and R designations. The terminals for three-phase motors are marked 1, 2, and 3. Service for the controls of these motors is performed in the same way as it is for any other motor.

12.9

Summary

Electric motors are used for many different purposes in a typical air conditioning or heat pump system. Usually, the evaporator blower, the condenser blower, and the compressor are all driven by electric motors. The motors and the controls for these units are important parts of any air conditioning or heat pump system.

Most of the motors used in air conditioning or heat pump systems are induction motors. These motors are relatively simple, but they develop sufficient power for the applications in which they are used. Different types of induction motors are used for different applications. The main differences among the motors are related to the starting characteristics or the running power developed by the motor. The main types of motors used are shaded pole motors, split-phase motors, permanent split-capacitor motors, capacitor start motors, capacitor start, capacitor run motors, and three-phase motors. The first five motors are all single-phase motors. The service checks for these motors are similar.

All induction motors operate in a similar way, but the methods used to start the motors vary with each type. All single-phase motors require some method of starting. Three-phase motors do not need a starting device because they have a naturally rotating magnetic field due to the characteristics of three-phase power.

The first check to make in servicing a motor is a check for mechanical failure. Then, the electric components must be checked. The electric components include the power supply, the overloads, the motor windings, the starting devices, and the capacitors.

The electric power supply and the overloads can be inspected visually for faults. Motor windings for single-phase and three-phase motors can be checked with an ohmmeter. Several service checks are required for the windings.

Starting switches are used to help start the motors. The most common devices are centrifugal switches, potential relays, and current relays. Some solid-state relays are used also for starting motors. Each type of starting switch has its own service procedure.

Both start and run capacitors are used in motor circuits for improving starting performance and running efficiency and power. These capacitors must be checked for faulty operation.

12.10

Questions

1. For what three purposes in an air conditioning system are electric motors used?
2. Name the two main causes of failure of electric motors.
3. Worn _____ are indicated if there is excessive movement when the motor shaft is moved by hand.

4. Name the five electric components of the motor that are directly related to its operation.
5. Is the following statement true or false? All motor overload protectors have automatic reset.
6. Name what the C, S, and R stand for on motor terminals.
7. When an ohmmeter is used to check the resistance through run and start windings, which set of windings should normally have the highest resistance?
8. A potential relay has a set of _____ and a _____ similar to those elements in a magnetic relay.
9. The coil in a potential relay is energized by voltage generated in the _____ windings.
10. A current-sensing relay is energized by the _____ in the run windings circuit.
11. Start capacitors are used to improve the _____ torque of a motor.
12. Run capacitors are used to improve the _____ of a motor, which makes the motor operate more efficiently.
13. Describe the electric motor installation in a hermetic compressor.
14. Is the following statement true or false? Hermetic compressor motors always use centrifugal switch starters.

Service Procedures Common to All Refrigeration Systems

13

13.1

Introduction

Many service procedures are common to all types of air conditioning and heat pump systems. Most of these procedures are related to the refrigeration system components. Regardless of the part of the system that is being checked, one or more of these service procedures may have to be performed. These common procedures are described in this chapter, and a discussion is given of the steps used in performing them.

Refrigerant is circulated through an air conditioning or heat pump system to carry heat from the low-pressure side of the system to the high-pressure side. Problems in the low-pressure side of the system affect the high-pressure side, and problems in the high-pressure side affect the low-pressure side. These effects in the system are usually shown by differences in pressure or in temperature of the refrigerant. Most service procedures that are common to all systems are related to the refrigerant pressures of an operating system.

This chapter covers procedures for checking refrigerant pressures in both sides of the system, checking the refrigerant charge, checking for leaks, pumping down a system, evacuating a system, and charging a system with refrigerant. Many of these procedures may have to be performed regardless of what part of the system is being checked. A service technician should be able to properly perform each check.

Checking Refrigerant Pressures

Various problems in an air conditioning or a heat pump system can cause high or low pressures. Higher-than-normal pressure on the high-pressure side of the system can be a result of an overcharge of refrigerant in the system, a reduction in the amount of air across the condenser coil, a dirty condenser coil, noncondensable gases in the condenser coil, or other problems. Low pressure may be caused by a low refrigerant charge in the system, a low condenser air temperature, or other problems.

The pressures on the high-pressure and the low-pressure sides of an air conditioning or heat pump system are checked by the use of a gauge manifold. The **gauge manifold** is a connecting manifold with valves, gauges, and hose connections arranged on it in such a way that it is easy to check pressure, evacuate a system, or add refrigerant to a system. The schematic of a typical manifold gauge set is illustrated in Figure 13–1.

A gauge manifold has a high-pressure gauge and a compound gauge on a common manifold. Each gauge

Gauge manifold

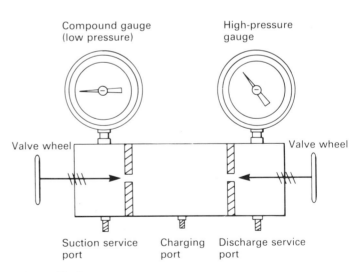

FIGURE 13–1
Schematic of manifold gauge set

is connected internally to a hose connection. The hose connections are used to run hoses from the manifold connections to the air conditioning or heat pump unit. Each gauge hose connection, in turn, is connected inside the manifold to a center hose connection port. The internal connection between each of the gauge ports and the center port is controlled by a valve on the manifold.

To use a gauge manifold to check the pressures in an operating air conditioning or heat pump system, turn the unit off. The hose from the high-pressure gauge port is connected to a service port on the high-pressure side of the system. The hose from the low-pressure gauge port is connected to a service port on the low-pressure side of the system. The service ports used for the connections can be on either backseated service valves or Schrader valves. If a hose is connected to a Schrader fitting, you must use either a Schrader fitting adapter or a hose with an adapter built into it. If service valves are used, they should be backseated before the connection is made.

Make sure that the valves on the manifold are closed so that refrigerant will not be lost from the center hose when the connections are made. Turn the service valve two turns inward to open the system to the service ports. When the gauges indicate that pressure is in each hose, purge the air out of the hose by slightly loosening the knurled nut connecting the hose to the manifold. When refrigerant flows from the loosened nut, close the nut again. Start up the unit, and observe the pressures indicated on the gauges in the two sides of the system.

13.3

Checking the Refrigerant Charge

Leaks in a system or improper charging may leave an air conditioning or a heat pump system with too little or too much refrigerant in it. Every system is designed to produce its rated cooling output at a given set of conditions with a specific refrigerant charge. Thus, a serviceperson must be able to properly check the charge and determine if it is the correct charge for the system.

There are several methods used to check the refrigerant charge in an air conditioning or a heat pump system. The best method to use for any given unit depends on the facilities for checking the charge that the manufacturer provided on the unit. The three main methods for checking the charge are the use of sight glasses, the use of liquid level valves, or the use of a pressure/temperature chart. The chart shows the operating pressures for various condensing air temperatures, and they are furnished by the equipment's manufacturer. Each method is described in the following subsections.

Sight Glasses

Sight glass

Some units are equipped with a **sight glass** in the liquid line where it leaves the condenser. To check a unit with the sight glass, turn on the unit. Let it run for 5 minutes or longer, until the condenser air temperature is about normal for the system. Then, observe the refrigerant in the sight glass.

With a proper refrigerant charge, or even an overcharge, the sight glass will be filled with liquid refrigerant. If bubbles appear in the refrigerant flowing through the glass, there is vapor in the refrigerant. The bubbles indicate that not all the refrigerant is condensing in the condensing coil. The bubbles are an indication of either an undercharge or some other problem in the condensing section that is preventing the refrigerant from condensing. If the problem proves to be an undercharge, then more refrigerant should be added to the system.

If an overcharge is suspected, purge refrigerant from the unit while it is running. Again, observe the sight glass. If bubbles start to appear in the glass, the unit is just slightly undercharged.

Liquid Level Test Valves

Liquid level indicators

Liquid level indicators are small valves, sometimes called snifter valves, that are installed on refrigerant lines, coil return bends, or liquid receiver tanks. The snifter valves have a very small outlet opening. When the valve is opened, a sample of refrigerant flows out. Figure 13–2 shows liquid level indicator valves for an evaporator coil.

To use a liquid level indicator valve to check the liquid level in a unit, open the valve slightly. Observe the refrigerant leaving the valve. If the refrigerant leaving the valve is in the gaseous state, then only refrigerant vapor is present in the system at the location of the valve. If liquid refrigerant leaves the valve, then the liquid level in the system is above the level of the valve.

If a unit has only one liquid level test valve, it is usually located on a return bend of the condenser coil or on the side of the receiver tank. It will always be located at a point in the system where the refrigerant should be in the liquid state, but just below the point where vapor may be present, for a given set of operating conditions. To use the valve for checking the liquid level in the unit, operate the unit at the normal operating conditions recommended by the manufacturer in the instruction manual. While the unit is operating, open the valve slightly in short intervals. If liquid only is emitted, the refrigerant level is above the valve. If vapor only is emitted, the refrigerant level is below the valve. If both liquid and vapor are emitted in spurts from the valve, the level of the refrigerant is just at the level of the valve. This level is just where it should be under normal operating conditions, and it indicates that the unit is properly charged.

If a system has two liquid test valves, they will be located at a high and a low position on the condenser coil return bends or on the side of the receiver tank. The lower valve is located on a part of the coil or tank where liquid should be present during normal operation. The upper valve is located where refrigerant vapor should be present during normal operation. To check the liquid level in the system, allow the unit to run long enough to reach stable conditions at normal operating temperatures. Then, open each valve in turn. There should be liquid refrigerant at the lower valve and vapor only at the upper valve. The liquid from the lower test port should be cold, and frost should form on the valve port while it is open. The vapor from the upper valve should be quite warm. If vapor is emitted from the lower valve, a low charge is indicated. If liquid is emitted from the upper port, an overcharge is indicated.

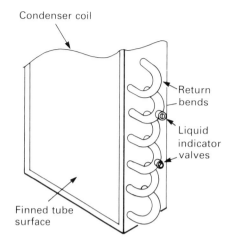

FIGURE 13–2
Liquid level indicator valves on evaporator coil

Pressure/Temperature Operating Charts

Pressure/temperature operating charts

Most manufacturers furnish **pressure/temperature operating charts** with their air conditioning or heat pump equipment. Such a chart is shown in Figure 13–3. The chart shows, in graph form, the normal pressures to be found in the high- and low-pressure sides of a system for various evaporator and condenser air temperatures.

To use one of these charts for checking refrigerant charge, operate the unit at typical operating conditions.

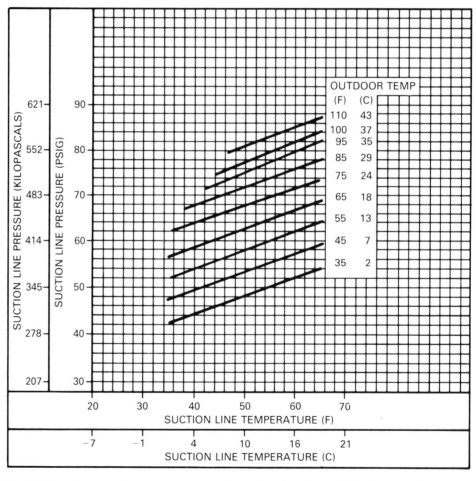

FIGURE 13–3   Pressure/temperature operating chart for air conditioning unit (Courtesy of Carrier Corporation)

Check the pressures in each side of the system with pressure gauges. The pressures read from the gauges are then compared with those shown on the graph for a properly operating system, at the same operating temperatures.

Another way to check the refrigerant charge with a pressure/temperature operating chart is to attach an accurate thermometer to the lower section of the condenser grille and measure the inlet air temperature. Slowly close the liquid line valve on the system until the low-side pressure is as shown on the chart for the refrigerant being used. Read the high-side pressure indicated on the chart for that particular low-side pressure. If the system is properly charged, the actual pressures and temperatures should match those shown on the chart.

13.4

Checking the System for Leaks

Regardless of which major section of an air conditioning or a heat pump system is being checked, you will often have to check that part for leaks. To check for leaks, you must pressurize part or all of the system with refrigerant. If the entire refrigerant charge has been lost from the system, pressurize the entire system, and search for the leak in the entire system. If the entire system needs to be pressurized, the best way to do it is through both the suction and the discharge service valves, as illustrated in Figure 13–4. By pressurizing through both valves at once, you apply pressure to both sides of the expansion device. If a unit only has a liquid line valve or a Schrader valve in the suction line, you may not be able to charge both sides at once. In this situation, charge through the available port, but make sure that refrigerant flows through the entire line system before checking for leaks.

Whenever service work is performed on the refrigeration system in an air conditioning or a heat pump system, there is a good chance that some of the refrigerant from the system will be lost. Also, if any leaks occur in the refrigerant lines in the system, some refrigerant is lost. Since the proper refrigerant charge is necessary

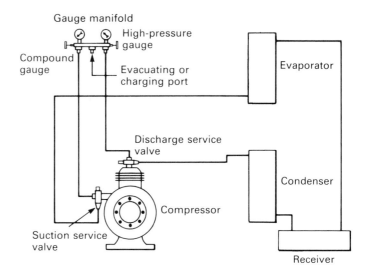

FIGURE 13–4
Method of attaching gauge set to air conditioning system for leak testing

if the unit is to operate the way it should, a service technician must know how to add refrigerant to a system. It is absolutely necessary that any refrigerant added to a system be dry and that no air be allowed to enter with the refrigerant. Since most refrigerants must be kept under pressure at normal temperatures, the flow of the refrigerant into the system is controlled by pressure. Thus, a service technician must also know how to charge a system properly when doing service work.

To apply pressure to the entire system for a system with both suction and discharge service valves on the compressor, backseat both service valves. Remove the service port caps, and attach the gauge manifold hoses to the suction and discharge service valve ports. Attach the hose from the center port of the gauge manifold to the drum of refrigerant you are using to charge the system, as shown in Figure 13–5. Turn both service valves to an intermediate position between open and closed. Open the drum valve, and open both valves on the gauge manifold. Opening these valves allows refrigerant to flow into the system. Do not try to pressurize the system beyond a pressure corresponding to the satu-

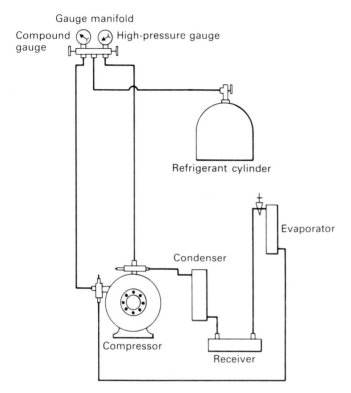

FIGURE 13–5
Connections for gauge set to charge system from refrigerant drum

ration pressure of the refrigerant for the existing ambient temperature.

Check for leaks with an electronic leak detector, if possible. If an electronic detector is not available, then a halide detector or soap bubbles can be used. Check all refrigerant lines and refrigerant line connectors in the system for leaks. Pass the detector probe, or place the soap solution, along the lines and around each connector. Most refrigerants are heavier than air, so it is important to check on the bottom of the joints as well as on the top. If an electronic detector is used, leaks will be indicated by the indicator needle or by an audible signal, depending on the type of detector used. If a halide detector is used, leaks will be indicated when the flame of the detector turns green. Bubbles will appear in the soap solution if soap is used as a detector.

When all leaks are found, make sure that the manifold valves are closed. Relieve the pressure in the system by removing the center hose from the refrigerant drum and releasing the refrigerant from the system into the atmosphere by opening the manifold valves. Repair the leaks, and evacuate the system before the recharging.

13.5

Pumping Down a System

Pumping down

An air conditioning or heat pump system with a liquid line valve and service valves on the compressor can be **pumped down** for service. This term means that the refrigerant charge is pumped into the condensing section of the system by the compressor. The liquid line valve and service valves on the system can then be used to isolate the low-pressure side of the system or the compressor. With the system pumped down, components and parts in the low-pressure side of the system or in the compressor can be worked on without losing the refrigerant charge from the system.

To pump down a system, wire a toggle switch into the control circuit so that unit operation can be controlled from the unit. Figure 13–6 shows the position of this switch in the control circuit. Turn the unit off with the toggle switch. Lower the setting on the low-pressure switch to zero. Turn the thermostat up high enough so that it will continuously call for cooling. Set the fan switch on the thermostat to the on position.

Now, attach a gauge manifold to the gauge ports on the two service valves on the compressor. Attach the high-pressure gauge to the discharge valve service port, and attach the low-pressure gauge to the suction service port. Open the service valve gauge ports by turning the valve stems one or two turns. Loosen the hose connection nuts one at a time to purge the air from the hoses, and then retighten the nuts. Close the liquid line valve by frontseating it. This valve is located on the outlet of

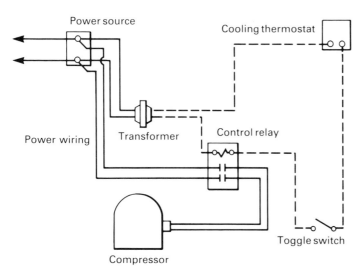

FIGURE 13–6
Toggle switch wired into control system

the receiver, if the system has a receiver. Or it is in the liquid line between the condenser and the filter drier.

Next, start the compressor by closing the toggle switch, and allow it to run until the suction pressure is reduced to near 0 pounds per square inch. Stop the compressor and observe the suction pressure. When the compressor stops, the pressure will normally go up somewhat, because of reexpansion of the refrigerant gas. It should then level off. Operate the compressor in cycles until the suction pressure remains at about 1 or 2 pounds per square inch when the compressor is not running. You may need to jumper the pressure switch to get it to come on when the pressure gets down close to 0 pounds per square inch. Always keep a slight positive pressure in the system to prevent air and moisture from entering, in case there is a leak in the system.

When the suction pressure remains at about 1 or 2 pounds per square inch, frontseat the discharge service valve on the compressor to isolate the refrigerant in the condensing section of the system. The refrigerant charge is now in the condensing section of the system between the discharge valve and the liquid line valve. Those parts

Temperature (°F)	Vacuum (in. Hg)
212	0.0000
205	4.9200
194	9.2300
176	15.9400
158	20.7200
140	24.0400
122	26.2800
104	27.7500
86	28.6700
80	28.9200
76	29.0200
72	29.1200
69	29.2200
64	29.3200
59	29.4200
53	29.5200
45	29.6200
32	29.7400
21	29.8200
6	29.8700
−24	29.9100
−35	29.9150
−60	29.9190
−70	29.9195
−90	29.9919

TABLE 13–1   Boiling Temperature of Water in Vacuum Range

of the system from which the refrigerant has been removed—from the liquid line valve to the discharge service valve—can now be worked on. If any part of the system is opened to make any repairs, that part should be evacuated before the refrigerant is allowed to enter that part of the system from the condensing section.

To place the unit in service, open the service valves. Slowly open the liquid line valve to allow the refrigerant to reenter the low-pressure side of the system. Reset the low-pressure switch to its normal operating pressure. Remove the toggle switch, and check the system for proper operation.

13.6

Evacuating a System

If a refrigeration system gets air in it, water vapor will probably get in also. To correct this condition, you must release the refrigerant and evacuate the system to dry it out. Then, you can recharge the system with clean, dry refrigerant. The evacuation process is intended to remove any water vapor that is in the system along with the old refrigerant.

To remove the water from the system, you must lower the pressure to the point where the water will vaporize at reasonable temperatures. This lowered pressure is a vacuum of 29.25 inches of mercury (abbreviated as in. Hg) at a temperature of 70 degrees Fahrenheit. Table 13–1 shows the vacuum necessary for moisture removal at different temperatures.

During the evacuation process, the pressure in the system should be measured with an accurate vacuum gauge—either an electronic gauge or a mercury manometer—while the evacuation process is performed. The evacuation pump used must be capable of pulling a low enough vacuum to vaporize the water at whatever the temperature is during the process.

To evacuate a refrigeration system, backset the service valves and attach a gauge manifold to the system. Attach the high-pressure gauge hose to the liquid line valve, and attach the low-pressure gauge hose on the suction side of the system. Use the valve port on the suction service valve for the low-pressure side of the system, or use a Schrader valve on the suction line. Install a vacuum gauge on the vacuum pump connection manifold. Connect the pump to the center port of the gauge manifold, as shown in Figure 13–7. Open the valves in the system and on the manifold, and start the pump.

Evacuate the system to at least 29 inches of mercury pressure. Observe the vacuum gauge on the pump during the operation. If the process takes an abnormally long time, there may be a leak in the system. Stop the pump at least once during the initial evacuation cycle to determine whether there is a rapid loss of vacuum because of a leak. If the system will not hold a vacuum

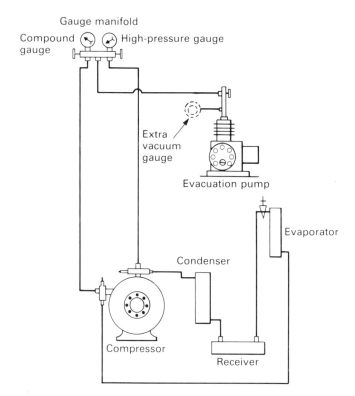

FIGURE 13–7
Gauge set connections for evacuating a system

with the pump off, there is a leak in the system. It must be repaired before you proceed.

The evacuation process should be continued until you are sure that all of the water in the system has been evacuated. When the evacuation process has been completed, close the manifold gauge valves, stop the pump, and disconnect it from the manifold. If the unit is to be charged, or if refrigerant is to be used to break the vacuum in the system, the center hose can now be attached to a drum of refrigerant, and the charging procedure followed.

Two or three cycles of evacuation and purging with clean refrigerant will normally be required to thoroughly clean and dry a contaminated system. To repeat the evacuation cycle, break the vacuum with clean refrigerant. That is, allow enough refrigerant to enter the system to raise the pressure a few pounds. Then, repeat the evacuation process. After the final cycle, the unit can be recharged.

13.7

Charging a System with Refrigerant

To add refrigerant to an air conditioning or a heat pump system, attach a manifold gauge set to the system. The hose from the high-pressure gauge should be attached to a service port on the high-pressure side of the system. The hose from the compound pressure gauge should be attached to a service port on the low-pressure side of the system. The center hose from the gauge set should be attached to the drum of refrigerant to be used for charging. See Figure 13–5 for the proper way to attach the service manifold.

Now, open the valve on the drum of refrigerant, and purge the hoses of any air they have in them. Close the valve on the refrigerant drum after the hoses are purged, and start the unit. Open the valve on the high-pressure gauge side of the manifold. This valve opens the high side of the unit to the center hose. The valve on the low-pressure gauge remains closed. Open the valve on the refrigerant drum to allow the refrigerant to flow into the unit.

Note: The pressure in the system may be greater than the pressure in the refrigerant drum. In this situation, the drum should be placed in a pan of hot water to increase the pressure so that the refrigerant will flow into the unit.

Watch the pressure gauges while the refrigerant is added to the unit. The refrigerant should be added in small quantities to prevent overcharging. Close the valve, and run the unit for several minutes after each addition of refrigerant in order to stabilize the operation of the unit. Check the operating pressures against the pressures shown for proper operation on the pressure/temperature chart. When the charge is correct, according to the chart, close the valve on the refrigerant drum. Back-seat the service valves on the unit, and remove the manifold gauge set.

13.8

Summary

Some service procedures are common to all parts of the air conditioning or heat pump system. That is, they may have to be performed regardless of what part of the system is being worked on. Most of these procedures relate to the refrigeration system and almost exclusively to the refrigerant in the system. A service technician must be able to perform these service procedures properly, since the refrigerant charge is very important to the function of a unit. These common procedures include checking the refrigerant pressures, checking the refrigerant charge, checking a unit for leaks, pumping down a system, evacuating a system, and charging a system with refrigerant.

The pressures on each side of an air conditioning or heat pump system are directly related to the operation of the system. Checking those pressures during operation is an important method used by servicepeople to determine how the unit is running. A manifold gauge set is used to check the pressures. The gauge set is also used in many other service procedures.

The correct refrigerant charge is necessary for proper operation of any air conditioning or heat pump system.

Improper service procedures or leaks in the system can cause a system to be overcharged or undercharged with refrigerant. A service technician must know how to check the refrigerant charge in a system.

Leaks in the refrigerant lines or connections on a system will allow refrigerant to leak out or allow air and water vapor to leak in. Thus, being able to properly detect and repair leaks is an important task of a service technician. Leaks can be detected with an electronic detector, a halide detector, or soap bubbles.

If repairs must be made in the low-pressure side of an air conditioning system, a heat pump system, or a compressor, the refrigerant charge can be pumped down into the condensing section of the unit. For this operation, the unit must have a liquid line service valve and service valves on the compressor. In many cases, pumping down a system saves the cost of new refrigerant when a unit requires service.

Evacuation of a system is always necessary if the system's refrigerant gets air or water vapor in it. Evacuation is the process of drawing a low enough vacuum on the system so that water in the refrigerant will vaporize and can be removed with the refrigerant gas. A good-quality vacuum pump and vacuum gauge are necessary for this process.

The proper refrigerant charge is necessary for an air conditioner or a heat pump to operate the way it should. Therefore, a service technician must know how to add refrigerant to a system. Any refrigerant added to a system must be dry, and no air can be allowed to enter with the refrigerant.

13.9

Questions

1. The _____ system used in an air conditioning system is similar to the one used in a heat pump system.
2. Describe a gauge manifold, and explain what it is used for.
3. How does a compound gauge differ from a high-pressure gauge?

4. Describe the sight glass used on a refrigerant line.
5. Is the following statement true or false? Liquid level indicators are located on a part of a refrigeration system where the refrigerant can be checked to see whether it is in the liquid or vapor state.
6. Describe a pressure/temperature operating chart.
7. Name the three most common methods used for testing for leaks in the refrigerant lines in a refrigeration system.
8. Is the following statement true or false? The pressure in a refrigeration system comes from air that is compressed by the system's compressor.
9. Describe the pumping down procedure used on a refrigeration system.
10. Describe the evacuation procedure used to get water vapor out of a refrigeration system.
11. Is the following statement true or false? A refrigeration system can be charged with any refrigerant.
12. Is the following statement true or false? A service gauge manifold is used in charging only as a convenience in getting the refrigerant into the system; it serves no other purpose.

14

Compressor and High-Pressure Side of the System

14.1

Introduction

The components and parts found in the high-pressure side of an air conditioning or a heat pump system, including the compressor, are described in this chapter. The compressor, condensing section, refrigerant lines, and refrigerant line accessories are described. The operation of each of these parts of the system is also explained. An air conditioning or heat pump technician must understand the operation of these components, how they function, and their relationship to each other in a system in order to be able to troubleshoot and service the system.

This chapter presents a description of the construction and the operation of the most common types of compressors used in air conditioning or heat pump systems. Air-cooled condensing sections are also covered. Condenser coils are discussed in detail. Condenser blowers are covered only briefly in this chapter; they are covered in detail in Chapter 18. The refrigerant lines for the high-pressure side of the system are discussed. The accessories commonly used in the refrigerant lines on the high-pressure side of the system are also included.

14.2

Types of Compressors

The **compressor** is the component that maintains a pressure differential between the high-pressure and low-pressure sides of an air conditioning or a heat pump system. Thus, it is one of the most important parts.

Various types of compressors are used in different systems. There are two general ways to classify compressors: by physical configuration and by general mechanical type. Each of these two classifications is covered in the subsections that follow.

Compressors Categorized by Physical Characteristics

There are three types of compressors with similar physical characteristics: open, semihermetic, and hermetic. Semihermetic compressors are also called serviceable hermetic compressors. These three types of compressors are illustrated in Figure 14–1. Each type is described in detail in the following subsections.

Open Compressors.   The open compressor, shown in Figure 14–2, is one in which the motor that drives the compressor is mounted externally to it. The connection between the motor shaft and the compressor shaft is either a direct connection or a belt and pulley. Open compressors are built so that they can be taken apart for service, and they are relatively simple in design.

One of the shortcomings of the open compressor is that the shaft must extend through the end of the compressor frame. Thus, the opening through which it passes must be sealed to prevent refrigerant and lubricating oil from leaking. Various types of seals are used around the shaft, but they are all subject to wear and thus require service at frequent intervals.

Open compressors

Semihermetic or Serviceable Hermetic Compressors.   A semihermetic, or serviceable hermetic, compressor is built with the motor and the compressor inside the same enclosure, as shown in Figure 14–3. An extension of the motor shaft is used as the crankshaft for the compressor. This arrangement eliminates the need for connecting devices between the motor and the compressor and hence eliminates the need for a shaft seal. This design also

Semihermetic, or serviceable hermetic, compressors

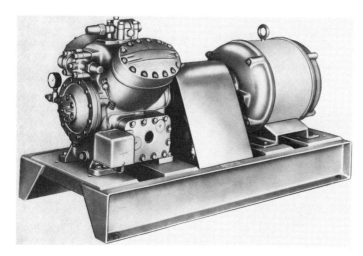

A. Open

B. Semi-hermetic

C. Hermetic

FIGURE 14–1
Three types of compressors used in air conditioning and heat pump systems (Courtesy of Copeland Corp. and Carrier Corp.)

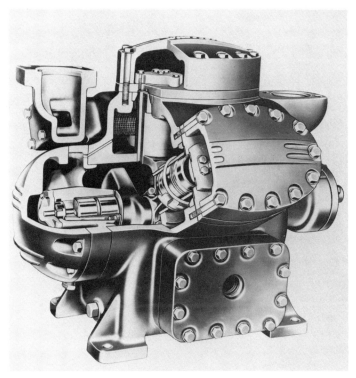

FIGURE 14–2
Cutaway view of open compressor (Courtesy of Carrier Corp.)

allows the refrigerant to be used as a coolant for the motor windings. Lubrication for both the motor and the compressor bearings is provided by oil from the compressor crankcase. A semihermetic compressor has removable heads, oil pan, and end bells so that it can be taken apart for service.

Semihermetic compressors are available in all sizes. But they are more common on air conditioning or heat pump systems that have at least 5 tons of cooling capacity. The original cost of a semihermetic compressor is greater than the cost of some of the other types. However, the fact that it can be repaired saves money in case of a breakdown.

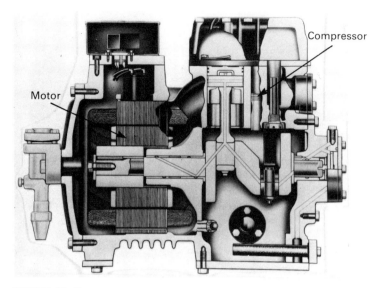

FIGURE 14–3
Cutaway view of semihermetic compressor (Courtesy of
Copeland Corp.)

Hermetic compressors

Hermetic Compressors.   A hermetic compressor, shown
in Figure 14–4, is one in which the motor and the compressor are built on the same frame. Both are sealed in
a common metal container during the manufacturing
process. A hermetic compressor cannot be opened for
service in the field. In case of a motor or a compressor
failure, the entire unit must be replaced.

Hermetic compressors are manufactured mostly in
small sizes, and they are used in small air conditioning
or heat pump units. Most hermetic compressors have
7 1/2 tons of cooling capacity or less, although some
larger ones are built. The motor in a hermetic compressor
is refrigerant-cooled. The lubricating oil for both the motor and the compressor is contained in the compressor
crankcase.

Compressors Categorized by Mechanical Action

The second method of categorizing compressors is by
mechanical action, that is, the way they actually com-

FIGURE 14–4
Hermetic compressor (Courtesy of Tecumseh Products Co.)

press the refrigerant in the system. There are four types
of compressors as defined by mechanical action: reciprocating, rotary, centrifugal, and screw. Each type is
described in the following subsections.

Reciprocating Compressors.   The reciprocating compressor is the most common type used in comfort air
conditioning or heat pump systems. It is the only com-

Reciprocating compressors

Positive-displacement compressor

pressor covered in the troubleshooting and service sections of this text. The other types are described in this section, but they are not covered in the service section.

A reciprocating compressor is a **positive-displacement compressor**. That is, as refrigerant gas goes through the compressor, it is forced into a space of a fixed volume. A positive-displacement compressor produces higher pressures than some of the other types of compressors. Mechanically, a reciprocating compressor has many parts similar to those found in an internal combustion engine. However, it operates in just the opposite way that an engine works. Instead of mechanical power being produced by the expansion of a gas, gas is compressed by the application of mechanical power.

A reciprocating compressor has one or more pistons. The pistons move back and forth in cylinders. The pistons are connected to a crankshaft by connecting rods. The crankshaft is turned by a motor, usually an electric motor. A back-and-forth, or reciprocating, motion of the pistons is produced by the crankshaft. Valves located above the pistons in the cylinder head control the flow of refrigerant into and out of the cylinder space. When the piston goes down in the cylinder, refrigerant gas flows into the cylinder through the suction valve from the suction line. When the piston moves up in the cylinder, the gas is forced out of the cylinder, through the discharge valve, and into the hot gas discharge line. Figure 14–5 gives an exploded view of the valves in the compressor.

The control of refrigerant pressure by a reciprocating compressor depends on the valves in the unit. As a piston moves down in its cylinder, the pressure of the gas in the cylinder is reduced. When the pressure gets lower than the pressure in the suction line, the suction valve opens. Then, refrigerant gas is drawn into the cylinder from the suction line. Figure 14–6 shows the piston during a suction stroke.

As the piston completes the suction stroke and starts to move up in the cylinder, the suction valve closes. The pressure in the cylinder thus increases. When the pressure in the cylinder is greater than the tension in the discharge valve plus the pressure in the discharge line, the valve opens. Refrigerant is forced out into the hot

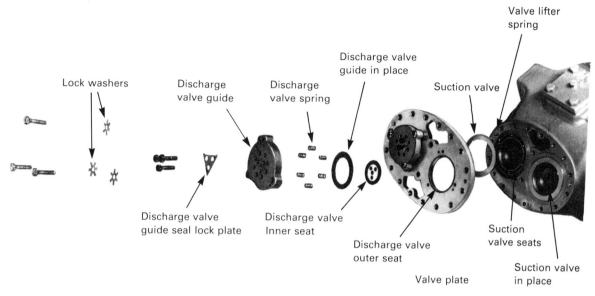

FIGURE 14–5   Exploded view of compressor head, valve plate, and valves (Courtesy of Carrier Corp.)

gas discharge line. Figure 14–7 shows the piston during a discharge stroke.

Two basic types of valves are used on compressors: leaf valves and plate valves. A **leaf valve** is a spring steel flapper fastened down at one end. It is held in place by the tension of the spring leaf itself or by other springs acting on the leaf. The **plate valve** has a floating plate held in place by small springs located around its perimeter. Figure 14–8 is an illustration of a valve plate for a plate valve. The spring tension in either type of valve holds it closed until the desired pressure is reached by the refrigerant in the system.

Leaf valve

Plate valve

Each piston in a compressor makes one suction stroke and one discharge stroke for every revolution of the crankshaft. The volumetric capacity of a compressor is based on cylinder volume, the revolutions per minute of the compressor crankshaft, and the number of cylinders in the compressor.

Reciprocating compressors are used in air conditioning or heat pump systems of all sizes. Since they are

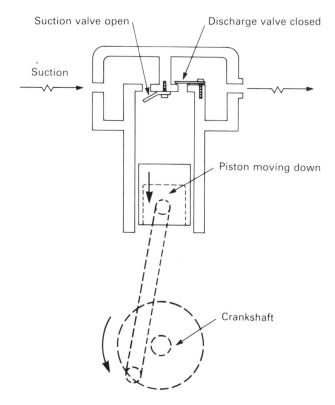

Suction valve open

Discharge valve closed

Suction

Piston moving down

Crankshaft

FIGURE 14–6
Piston during suction stroke

positive-displacement compressors, they operate efficiently in systems requiring relatively high pressures.
This type of compressor is adaptable for use with many
different refrigerants.

Rotary compressors

Rotary Compressors.   The rotary compressor receives
its name from the way it works. A rotary compressor
has a cylindrical ring that rotates, or turns, inside a round
cylinder. The ring is somewhat smaller than the cylinder. It is located so that it is in contact with the cylinder
at one point at all times. The ring is mounted off-center,
or eccentrically, in relation to the shaft. As the shaft
turns, one point on the ring's perimeter stays in contact

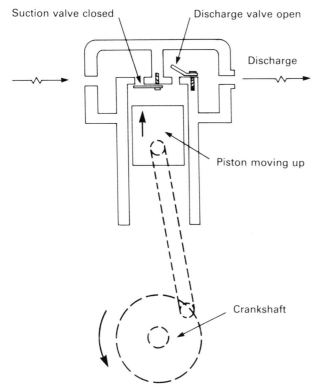

Suction valve closed   Discharge valve open

Discharge

Piston moving up

Crankshaft

FIGURE 14–7
Piston during discharge stroke

with the wall. A crescent-shaped opening is left between the ring and the wall of the cylinder opposite the side where they touch, as shown in Figure 14–9. The ring is turned by an electric motor. As the ring turns, the crescent-shaped opening moves around the cylinder wall. As the opening moves around the cylinder, refrigerant is drawn into it at one point and forced out at another.

There are two general types of rotary compressors: sliding vane compressors and rotating vane compressors. The **sliding vane compressor** has a spring-loaded vane in the side of the cylinder wall. This vane rides against the ring as it turns and forms a seal. As the ring turns, a low-pressure area is created in the crescent-shaped opening as it moves away from the sliding vane. A high-pressure area is created as it moves toward the

Sliding vane compressor

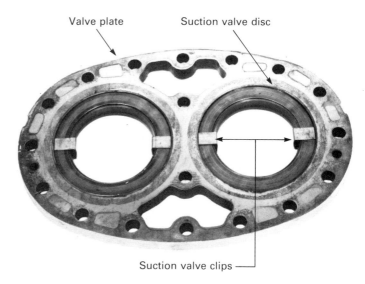

FIGURE 14–8
Compressor valve plate (Courtesy of Carrier Corp.)

vane. Figure 14–10A shows the beginning of the stroke. Figure 14–10B shows how suction and compression start on opposite sides of the sliding vane. Figure 14–10C shows the point at which the suction and compression are equal. And Figure 14–10D shows the point at which maximum suction and compression are obtained.

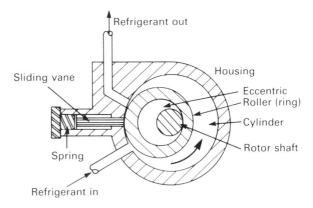

FIGURE 14–9
Schematic of rotary compressor

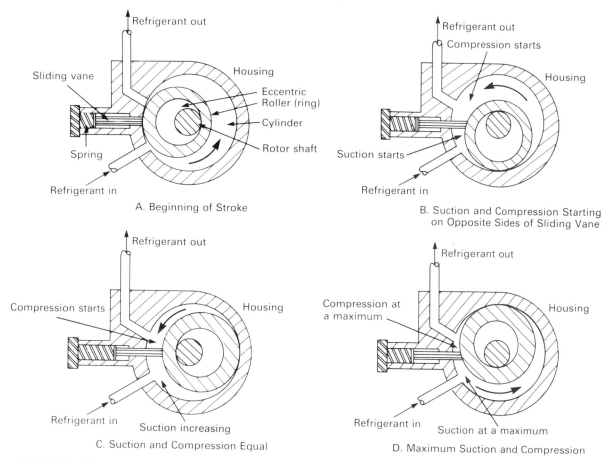

FIGURE 14–10   Schematic showing operation of sliding vane compressor

Openings in the sidewall of the cylinder are connected to the suction line of the system on the side of the vane where the low-pressure area occurs. Openings in the sidewall are also connected to the discharge line of the system on the side where the high pressure occurs.

A **rotating vane compressor** works on the same basic principle as the sliding vane compressor. The difference is that the rotating vane compressor has multiple vanes located in the ring, as illustrated in Figure 14–11. These vanes are forced against the cylinder wall to form a seal. As the ring rotates, refrigerant is drawn into the cres-

Rotating vane compressor

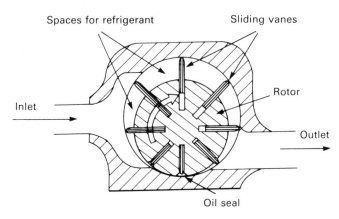

Spaces for refrigerant   Sliding vanes

Inlet

Rotor

Outlet

Oil seal

FIGURE 14–11
Rotary vane compressor

cent-shaped openings between the ring and the cylinder
wall, just as it is in the sliding vane compressor. The
suction line and hot gas discharge line of the system are
connected to the compressor in such a way that refrigerant is drawn in, compressed, and discharged at a higher
pressure as the cylinder rotates.

Rotary compressors are positive-displacement
compressors and operate very efficiently. Clearance between the moving parts has to be very close, so the rings
and cylinders are precisely machined. The actual seal
between the surfaces is the thin film of lubricating oil
used in the system. Rotary compressors are used mostly
in small air conditioning or heat pump systems, such as
window air conditioning or heat pump units and small
packaged units.

Centrifugal compressors

Centrifugal Compressors.   A centrifugal compressor,
shown in Figure 14–12, works on the same principle as
a centrifugal air blower. An impeller wheel with many
blades on it turns inside a housing. An opening in the
side of the housing, near the center of the wheel, is the
intake for the compressor. An opening on the outer perimeter of the housing is the outlet. The impeller wheel
turns at a high rate of speed. Refrigerant is drawn into
the wheel through the intake opening. The pressure of
the refrigerant increases as it goes through the vanes
on the wheel. It is finally thrown off the tips of the wheel

FIGURE 14–12
Centrifugal compressor (Courtesy of Carrier Corp.)

at a high velocity and a high pressure by centrifugal force. The high-pressure refrigerant leaves the compressor housing through the discharge opening.

A centrifugal compressor is not a positive-displacement compressor. It has the ability to move a large volume of refrigerant, but does it at a relatively low pressure. Its ability to move large volumes of refrigerant makes it a good compressor to use in large air conditioning or heat pump systems. The low pressure difference between the entering and leaving refrigerant can be compensated for by a refrigerant that boils at fairly low pressures. Centrifugal compressors using R–11 refrigerant are often found on large, multiton systems.

If higher pressures are needed for a particular application, a **multistage compressor** can be used. A multistage compressor has multiple impeller wheels. It is designed in such a way that as the refrigerant leaves the

Multistage compressor

outer rim of one wheel, it is directed to the center of the next. With this design, the pressure is raised in stages as the refrigerant goes through each wheel. This type of compressor can move large volumes of refrigerant at relatively high pressures.

Screw compressors

Screw Compressors.   In the screw compressor, shown in Figure 14–13, two screwlike sets of helical gears, contained in a tight-fitting metal box, interlock as they turn. The gears are turned by an electric motor. As the gears turn, refrigerant is drawn into the box, is compressed by being forced into a small space, and is discharged out the other end. Figure 14–14 shows the operating parts of a screw compressor. There is a suction inlet at one end of the gear box, and an outlet is located on the other end.

A screw compressor is a positive-displacement compressor and is very efficient. The parts of this compressor must be machined to very close tolerances. Also, proper lubrication is important in maintaining a seal between the moving parts. Screw compressors are usually used on large air conditioning or heat pump systems.

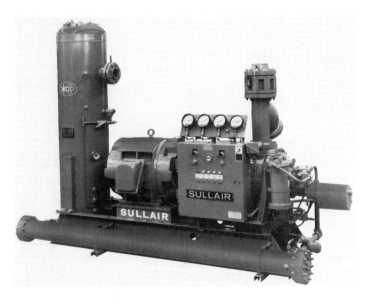

FIGURE 14–13
Screw compressor (Courtesy of Sullair Corp.)

Gears

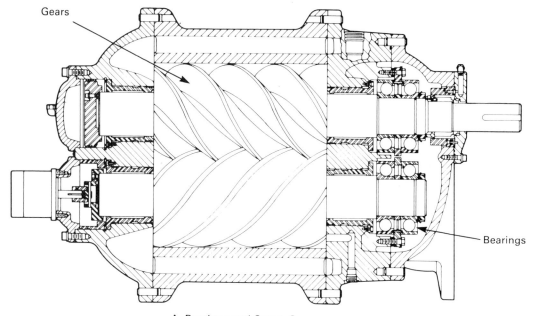

Bearings

A. Bearings and Screw Gears

Valve assembly

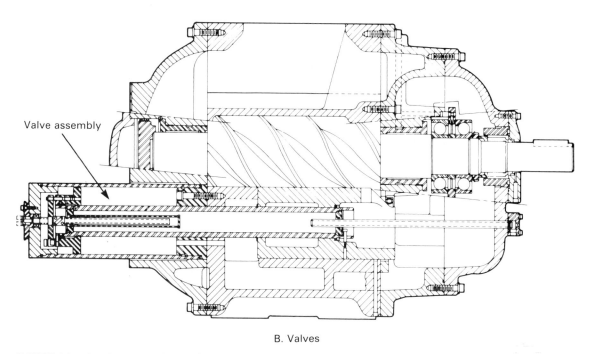

B. Valves

FIGURE 14–14   Cutaway views of screw compressor (Courtesy of Sullair Corp.)

Condenser

Subcooling

FIGURE 14–15
Air-cooled condensing section
(Courtesy of Rheem Co.)

Condenser Section

A **condenser** is basically a heat exchanger in which heat from the refrigerant is given off to the condensing medium. The heat given off is the heat picked up from the evaporator and the compressor. The most common condensing mediums for air conditioning or heat pump systems are air or water. This text covers only air-cooled condensing systems.

As refrigerant gas leaves the compressor, it is at a high temperature. As this gas passes through the condenser coil, its temperature is greater than the temperature of the air going across the outside of the coil. So, heat passes from the refrigerant to the air.

The most common type of air-cooled condenser is a finned-tube coil with a blower for moving air across it. Most condensers have propellor blowers. The condenser section of an air conditioning or a heat pump system is designed for an outdoor location, where outdoor air can be used for the condensing medium. Figure 14–15 shows a typical air-cooled condensing unit. The condenser section of an air conditioning or a heat pump system must be sized so that its condensing capacity matches the capacity of the other components in the system.

In operation, when the refrigerant in a system leaves the compressor, it is at a superheated temperature. Thus, the temperature of the refrigerant is higher than the saturated vapor temperature for the pressure. Most of the superheat is lost as the refrigerant moves through the hot gas discharge line. The refrigerant then enters the condenser coil as a saturated vapor.

In the coil, heat flows from the refrigerant to the condenser air because of the temperature difference between the two. As the heat is removed, the refrigerant condenses to a liquid. The condenser coil is sized so that all of the refrigerant will condense into a liquid before it leaves the coil. The last few rows of the condenser coil are used to subcool the refrigerant. **Subcooling** means that the refrigerant is at a lower temperature than the saturated temperature for the refrigerant at a given pressure. The refrigerant leaves the condenser section as a high-pressure, subcooled liquid. Figure 14–16 illustrates some typical temperatures for the heat transfer process in a condenser.

14.4

Refrigerant Lines

There are two refrigerant lines on the high-pressure side of an air conditioning or a heat pump system. One line is the **hot gas discharge line**, running from the compressor to the condensing unit. The second line is the **liquid line**, running from the condensing unit to the expansion device. These two lines are shown in Figure 14–17.

Hot gas discharge line

Liquid line

14.5

Accessories

Some of the accessories in the high-pressure side of an air conditioning or a heat pump system are service valves, sight glasses, moisture indicators, and liquid receivers. Service valves are actually found in both the low- and the high-pressure side of a system, but they will be cov-

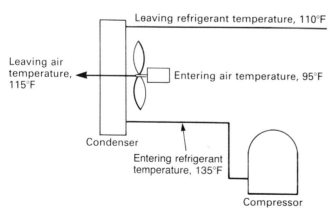

FIGURE 14–16
Heat transfer from refrigerant in condenser coil to air

Service valves

Valve wrench

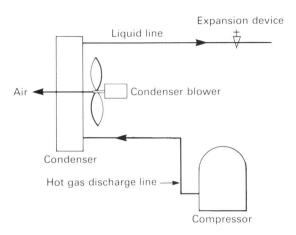

FIGURE 14–17
Hot gas discharge line and liquid line

FIGURE 14–18
Service valve (Courtesy of Primore Sales Inc.)

ered here. These accessories are all located in the refrigerant lines of the system. Each device is described in the following subsections.

Service Valves

On most systems, and especially on those with serviceable compressors, **service valves** are located on the side of the compressor where the hot gas discharge line leaves the compressor. They are also located on the suction side where the suction line enters the compressor. Some systems, especially those with hermetic compressors, have a single service valve located on the liquid line. Service valves may be used to shut off the refrigerant on either the suction or the discharge side of the compressor in order to isolate the compressor from the rest of the system. But most commonly, they are used for attaching pressure gauges to the system for various service procedures.

Figure 14–18 shows a typical service valve used for air conditioning and heat pump systems. A service valve is a hand valve having a stem with very fine threads. The stem is turned in or out by the use of a special wrench called a **valve wrench**, which is a ratchet wrench built especially to fit the valve stem. A service valve has an inlet port and two outlet ports. The **inlet port** is for connection to the part of the system from which refrigerant will flow. The **main outlet port** is for connection to the part of the system to which refrigerant will flow. The other outlet port is called the **service port**, and it is used for connecting a hose to the gauge manifold or for other service procedures.

When the valve stem is turned all the way in, the main inlet port of the valve is closed. The stem closes against a valve seat in the front of the valve, called a **frontseat**. See the cutaway view of a service valve in Figure 14–19. When the stem is turned all the way out,

Inlet port

Main outlet port

Service port

Frontseat

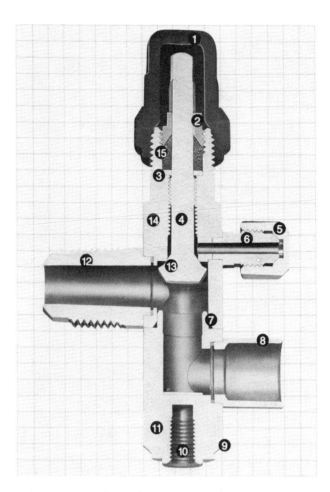

TYPICAL VALVE
ASSEMBLY

1. Valox Stem Cap
2. Packing Gland Nut
3. Packing Washer
4. Stem
5. Seal Cap
6. ¼ Flare Gage Port
7. Front Seat
8. Side Body (Sweat Fitting)
9. Coined Square
10. Bolt Hole
11. Lower Body
12. Side Body
   (Flare Fitting)
13. Back Seat
14. Upper Body
15. Packing

FIGURE 14–19   Cutaway view of service valve (Courtesy of Primore Sales Inc.)

Backseat

Sight glasses

Moisture indicator

it closes the service port by closing against a seat at the back of the valve, called the **backseat**. When the valve is in between the frontseat and the backseat positions, all three ports are open.

During normal operation of an air conditioning or a heat pump system, the service valves are in the backseat position. Thus, the system line is open and the service port is closed. To open the service port, turn the valve to an intermediate position. When the valve is frontseated, it shuts off the line the valve is located in.

Service valves have caps to cover the valve stem and service port opening when they are not being used for service. These caps have sealing gaskets and should always be left in place when the valve is not being used. The cap prevents the leakage of refrigerant from around the valve stem or the packing.

Sight Glasses

Sight glasses are installed in the refrigerant lines in a system to allow an operator or service technician to observe the flow of refrigerant in the line. A sight glass is a fitting that has a glass window in it, as shown in Figure 14–20. The window allows the operator to see the refrigerant in the line.

A sight glass in the hot gas discharge line just ahead of the condenser coil will normally show only refrigerant gas flowing, with a small quantity of oil in it. A sight glass in the liquid line leaving the condenser coil will show all liquid in the line. Bubbles or foam in the refrigerant in a sight glass indicates a shortage of refrigerant or some other problem in the system.

Moisture Indicators

Moisture indicators are often incorporated in sight glasses. The **moisture indicator** is a patch of colored chemical inside the glass window of the sight glass. The color of the chemical patch indicates whether or not there is water vapor present in the refrigerant. In a dry system—that is, one with no water vapor in the refrigerant—the chemical patch will be a particular color. If there is water vapor present in the refrigerant, the color of the patch will change.

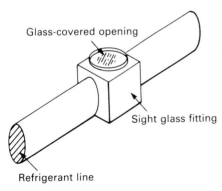

Glass-covered opening

Sight glass fitting

Refrigerant line

FIGURE 14–20
Sight glass

Different colors for moisture indicators are used by different manufacturers. To find the proper colors for wet and dry conditions on any particular sight glass, look for a note on the device. Or check the manufacturer's literature that comes with the sight glass.

Liquid Receivers

A **liquid receiver** is a tank installed in the liquid line of an air conditioning or a heat pump system. The receiver is used to store liquid refrigerant during the operation of the system. The receiver accommodates any changes in volume of the refrigerant in the system as the cooling or heating load changes or as the condensing air temperature changes. If the cooling load goes up or down, a greater or smaller part of the evaporator coil is used to evaporate the refrigerant. And the volume of the refrigerant in the system changes as the ratio of liquid to vapor changes. If the condensing air temperature goes up or down, more or less refrigerant is condensed as it goes through the condenser coil. Again, the volume of the refrigerant in the system changes. The variations in the volume of the refrigerant are compensated for by the storage capacity of the receiver.

A liquid receiver is constructed in the following way. As refrigerant flows into it, the liquid naturally lays in the bottom of the receiver, and vapor rises to the top. See the cutaway view of a receiver in Figure 14–21. The refrigerant line from the receiver tank to the expansion valve comes from the bottom of the tank, so only liquid refrigerant flows out of the tank. Use of a liquid receiver in an air conditioning or a heat pump system ensures that there will be liquid refrigerant in the refrigerant line ahead of the expansion valve under all operating conditions.

Liquid receiver

14.6

Summary

The compressor, the condenser, the hot gas discharge line, the liquid line, and various accessories associated with these lines are all part of the high-pressure side of

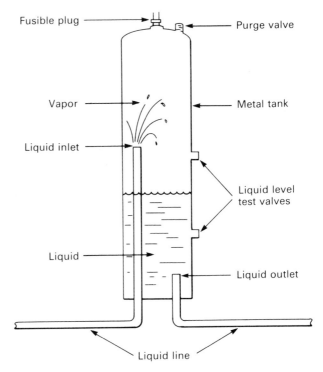

FIGURE 14-21
Cutaway view of liquid receiver

an air conditioning or a heat pump system. Each part is important to the operation of the system. So, an air conditioning or heat pump technician must understand the operation of each of these parts and know how they relate to each other in order to properly troubleshoot or service a system.

There are three basic physical types of compressors: open, semihermetic, and hermetic. An open compressor is one in which the motor that drives the compressor is separate from it. The motor is connected to the compressor either directly or by pulleys and a belt. The semi-hermetic compressor has its motor in the same housing as the compressor, but the housing can be taken apart for service. For a hermetic compressor, the motor and the compressor are in the same housing. The housing is sealed so that it cannot be opened for service.

Four major types of compressors are commonly used in air conditioning and heat pump systems: reciprocating, rotary, centrifugal, and screw compressors. Each type of compressor has special characteristics that make it best for certain applications. Reciprocating compressors are the most commonly used compressors.

The function of the compressor in any air conditioning or heat pump system is the same, regardless of the type used. The compressor's function is to raise the pressure and the temperature of the refrigerant so that heat can readily be removed from it by the condensing medium. A refrigerant line, called a hot gas discharge line, carries the high-pressure, hot refrigerant gas from the compressor to the condensing section of the system.

A second major part of the high-pressure side of an air conditioning or a heat pump system is the condensing section. The condenser is the part of the system in which heat that was picked up in the evaporator is rejected. Air-cooled condensing systems include a condensing coil and a blower. Hot refrigerant gas from the compressor passes through the coil, and air is blown over the coil. Heat in the refrigerant transfers to the air because of the temperature difference between the two. The refrigerant condenses and becomes a liquid as the heat is removed. The liquid refrigerant leaves the condensing section and flows to the expansion device through the liquid line.

Some of the most common accessories used in the high-pressure side of an air conditioning or heat pump system are service valves, sight glasses, moisture indicators, and liquid receivers. All of these parts and accessories are important to the operation of an air conditioning system. A malfunction in any part can cause an entire system to be shut down.

14.7

Questions

1. Name the three types of compressors, as categorized by physical characteristics, used in air conditioning and heat pump systems.

2. What is the greatest shortcoming of an open compressor?

3. A semihermetic compressor _____ (can, cannot) be dismantled and repaired in case of breakdown.

4. Is the following statement true or false? A hermetic compressor cannot be repaired in the field.

5. Name the four types of compressors as categorized by mechanical action.

6. Name the type of compresssor that has a crankshaft, cylinders, and pistons.

7. Discharge and suction pressure is controlled by the _____ in a compressor.

8. Name the two basic types of valves used in refrigeration compressors.

9. Describe the mechanical action of a rotary compressor.

10. Name the two general types of rotary compressors.

11. Describe the differences in the vanes in the two types of rotary compressors.

12. The seal between the moving parts of any compressor is the thin film of _____ oil.

13. A centrifugal compressor works on the same principle as a centrifugal blower; that is, an _____ rotates inside the compressor housing.

14. A centrifugal compressor moves a large _____ of refrigerant, but it does so at a fairly low _____.

15. A compressor that has two helical gears that mesh to compress the refrigerant is called a _____ compressor.

16. Describe a typical condenser used on an air conditioning system.

17. What is meant by *subcooling?*

18. When a service valve is frontseated, which port is closed?

19. When a service valve is backseated, which port is closed?

20. Why is a sight glass used in a refrigeration system?

21. What changes the color of the chemical patch in a moisture indicator?

22. What is the function of a liquid receiver in a system?

Servicing the High-Pressure Side of the System

15.1

Introduction

This chapter describes specific procedures for trouble-shooting or servicing components and parts found in the high-pressure side of an air conditioning or a heat pump system. The compressor is one of the components of the high-pressure side of the system. There are several different types of compressors available, and different compressors are used in different types of air conditioning and heat pump systems. The type most commonly used is the reciprocating compressor. Service procedures for a reciprocating compressor are discussed in this chapter.

There are two types of condensing units commonly used in air conditioning or heat pump systems, air-cooled and water-cooled. Air-cooled condensing systems are the most common type found in small comfort systems. And they are the only systems covered in this text. Service for air-cooled condensing systems is discussed in this chapter. Service procedures for problems occurring in the refrigerant lines on the high-pressure side of the system and in the accessories found in those lines are also described.

By following the procedures covered in this chapter, a service technician will be able to identify specific faults related to the parts in an air conditioning or a heat pump system. By repairing or replacing the faulty parts, the technician will be able to make the system operable.

Compressor

Reciprocating Compressors

A **compressor** is a mechanical device used in air conditioning or heat pump systems to compress refrigerant vapor. It has many mechanical parts. Failure of any of these individual parts will prevent the compressor from doing its job.

To service a reciprocating compressor, you must check the performance of all the parts to determine whether they are operating as they should. For convenience in describing the procedures for servicing a compressor, we can think of the compressor as being made up of three different sections. The first section, called the mechanical section, includes the crankshaft, connecting rods, and pistons. The second section includes the valve plate and valves. The third section is the lubrication system of the compressor. Figure 15–1 is a cutaway view of a compressor showing the parts related to the three categories.

An open or semihermetic compressor can be taken apart for service, and each of the parts can be checked. This dismantling and checking can be done in the field, but it is much more practical to do it in a shop. A hermetic compressor is nonserviceable and must be replaced if something is mechanically wrong with it.

The following subsections describe the service procedures for the three compressor sections.

Mechanical Parts

Seizing up

Mechanical breakage in the crank mechanism of a compressor is usually indicated by the compressor's **seizing up**, or refusing to turn. The motor stalls, and the overloads or fuses will open. A clamp-on ammeter can be placed around one of the power leads to the motor, and the motor can be energized. A high amperage reading will show on the ammeter until the fuses blow or the overloads open. If an amperage check indicates mechanical breakage, an open or semihermetic compressor can be dismantled and inspected.

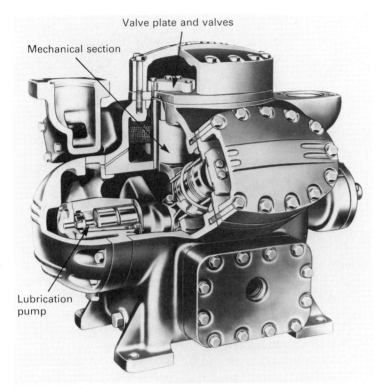

Valve plate and valves

Mechanical section

Lubrication pump

FIGURE 15–1
Cutaway of compressor showing internal sections (Courtesy of Carrier Corp.)

Dismantling.   To disassemble an open or semihermetic compressor, make sure that the electric power is turned off at the unit disconnect. Remove the power and control wiring from the compressor. Frontseat the service valves to isolate the compressor from the system, and disconnect the refrigerant lines. Note, or mark, the location of the head (or heads), end plates, and pan on the block of the compressor so that each can be replaced in its original position when repairs are finished.

Remove the head bolts. Most compressor heads are dowelled in place. So, loosen the head from the valve plate by placing the blade of a screwdriver under the head. Next, move the screwdriver around the head while tapping with a hammer until the gasket is loosened.

Then, lift the head straight up off the dowels. The discharge valves can now be seen on the top of the valve plate or on the cylinder block. Do not disturb the valves at this time, but remove the valve plate, if there is one. The piston heads can now be seen in the cylinder openings.

Drain the oil out of the crankcase through the oil inspection port. Be careful in draining and disposing of the oil, since it may be acidic. Remove the compressor end bells and the drain pan by using the same procedure given for removing the head.

Inspection.   Visual inspection at this point should show if there is any physical breakage of the parts in the crank assembly of the compressor. If there is any damage, the broken parts should be removed.

The method for removing the crankshaft depends on the type of crankshaft and connecting rods used in the system. If a regular crank type of crankshaft is used, and if the pistons have articulated connecting rods that move back and forth in wrist pins at each end, then the connecting rod caps can be removed from the bottom. The pistons can be pushed up in the cylinders. The shaft should now slide out the end of the compressor. If a concentric-ring crankshaft is used, and if the connecting rods do not have removable caps on the bottom, then the crankshaft will have to be worked out through the rings on the bottoms of the connecting rods as the pistons are manipulated. After the crankshaft has been removed, the pistons and the connecting rods can be removed from the cylinders.

If you are working on a semihermetic compressor, you can now remove and inspect the electric motor parts, also. All of the compressor parts should be cleaned in solvent and inspected for damage. Any broken parts should be replaced with identical replacement parts. The block should be completely cleaned both inside and out before reassembly begins.

Reassembly.   To reassemble the compressor, reverse the procedures outlined for dismantling. Use all new gaskets as the plates are replaced. Fill the crankcase with the proper charge of refrigerant oil, as specified by the compressor's manufacturer. Reconnect the refrigerant lines, electric power, and control wiring. Open the service valves, and thoroughly evacuate the system. Recharge the system with the proper refrigerant, as recommended by the unit's manufacturer. Start up and test-run the compressor, checking the pressures in the high- and low-pressure sides of the system to be sure it is operating properly.

Suction and Discharge Valves

A compressor may operate after a call from the thermostat for cooling or heating, but there may be little or no cooling effect. In addition, the pressure gauges may show no pressure difference between the low-pressure and high-pressure sides of the system. In both situations, the compressor valves may be broken. The compressor valves should be checked for breakage and replaced if they are faulty.

Checking for Broken Valves.   Most open or semihermetic compressors have service valves on both the suction inlet and the discharge outlet of the compressor. To check a compressor with two service valves for broken internal valves, make sure that the suction and discharge service valves are backseated. Attach the hoses of the gauge manifold to the service ports on the valves. The proper way to connect the gauges is shown in Figure 15–2.

Now, frontseat the suction service valve. Open the discharge valve stem one to two turns to open the service port to the gauge. Jumper the low-pressure control contacts so that the unit will not shut off on low pressure. Start the compressor. Run it until the low-pressure gauge has a reading in the vacuum range. If the pressure will not go down into the vacuum range, the suction valves are leaking. Shut off the unit, and observe the pressure gauges. If the pressure increases rapidly when the compressor is turned back on, then the discharge valves are leaking.

Replacement.   If the valves are leaking, they can be replaced by using the following procedure. Frontseat both service valves to isolate the compressor from the

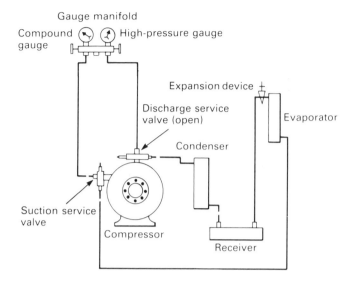

FIGURE 15–2
Attaching service gauge manifold to system

system. Disconnect the refrigerant lines from the compressor. Remove the head bolts and the head from the compressor. The discharge valves can then be seen on the top of the valve plate, if the compressor has one, or on the top of the compressor block. If there are any broken parts located on the top of the valve plate, remove them. Clean up any metal filings before lifting the valve plate.

Remove the valve plate, and inspect the suction valves that are located on its underside. Again, if there are any broken parts, keep them from falling into the cylinder openings. Examine the cylinder openings for any broken parts. Check the piston heads and cylinder walls for damage from broken parts. Clean out the cylinder opening and piston heads carefully.

Dismantle the valves by removing any fasteners that hold them. Clean off all surfaces. Inspect the valve seat to make sure that it is not damaged. If the valve seat is damaged in any way, it should be replaced with a new one. Replace all valves with new valves that are identical to the valves removed. Clean off all old gasket material from the compressor block, valve plate, and head.

Reassembly.   Install new gaskets, and replace the valve plate and head in the same positions that they were in originally. Reinstall head bolts. Reconnect the refrigerant lines to the compressor, and backseat the service valves. Reconnect the electric power and control lines to the compressor. Start up and operate the compressor until the system is stabilized. Check the operation by observing the high-side and the low-side pressures.

Lubrication

Lubrication of the compressor and the motor in a semihermetic or hermetic compressor is done by the lubricating oil contained in the crankcase. Only special refrigeration system oil should be used for this purpose. **Refrigeration oil** is a lubricant with a nonparaffin base that will perform well at low temperatures. It is also processed so that it is clean and free of any water in solution. A clue about problems within a system can come from the lubricating oil.

Refrigeration oil

Problems Indicated by the Oil.   Most refrigerants will pick up some oil vapor and carry it with them through the system. If the piping system is not engineered properly, this oil can settle out in the coils or other parts of the piping system and not be available to lubricate the compressor. Thus, the oil level in the compressor crankcase can be an indication of improper piping design. The oil level should be checked on a regular basis. If oil has to be added often, the design of the refrigeration piping system should be checked to make sure that the oil is not migrating out into the system and staying there.

If the motor in a semihermetic or a hermetic compressor overheats, the oil breaks down chemically. Excessive heat causes oil and refrigerant vapor to form acids. These acids damage the motor insulation. In extreme cases, they corrode the copper lines in the system and deposit the copper on the metal parts inside the compressor. Acidity of the oil is an indication of how the compressor has been operating.

A check for oil level and oil acidity involves a check of the oil reservoir, the oil pump, and the sight glass. These topics are covered next.

Oil reservoir

Centrifugal oil pump

Oil sight glass

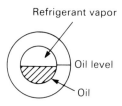

Refrigerant vapor

Oil level

Oil

FIGURE 15–3
Oil level sight glass

Oil Reservoir.   The oil reservoir on most small compressors—up to 20 tons of cooling—is located in the crankcase of the compressor. An oil pump is built into the compressor crankshaft. On large compressors, the oil reservoir may be in a separate tank. And there may be a separate pump for circulating oil to the bearings and other surfaces needing lubrication.

Centrifugal Oil Pump.   A centrifugal oil pump is used on most small compressors for comfort air conditioning or heat pump systems. This pump is an integral part of the crankshaft assembly. It works the same way on either horizontal or vertical shaft compressors. The shaft has a small-diameter hole drilled through the center in its long axis. The bearing on the oil inlet end of the shaft is constructed so that oil can get to the hole at the end of the shaft. Holes are also drilled in the crank arms of the shaft. Connecting holes are drilled in the connecting rods and wrist pins. The holes in the crank arms and connecting rods are at a right angle to the hole in the shaft. As the shaft rotates, oil is drawn into the hole in the shaft by capillary action. The holes at the ends of the rods allow the oil to flow through the wrist pins to lubricate the bearing surfaces there. Lubrication is also provided by oil that is splashed onto the cylinder walls as the crankshaft moves through the oil in the oil pan.

Oil Sight Glass.   Many compressors have an oil sight glass in the side of the crankcase housing. Figure 15–3 illustrates a sight glass. The sight glass allows the operator to observe the oil level. The sight glass usually has a mark etched on it at about the middle. This line indicates the proper oil level when the unit is not running. While the unit is running, the oil level can vary from the bottom of the glass to the top, but it should be visible in the glass. The level is usually higher in the glass when the unit is first shut off after operating because of refrigerant in the oil. On start-up, the oil level may drop momentarily, but it should then be seen somewhere in the glass as unit operation stabilizes.

Checking the Oil Level.   With or without a sight glass, a unit should have an oil level plug in the base. If there is no sight glass, this plug must be removed to check the oil level. *Note:* The plug should only be removed when the unit is not operating and when the pressure in the crankcase has been reduced to a minimum.

However, before removing the plug, you must reduce the crankcase pressure to atmospheric pressure. This task is done by closing the suction service valve and running the unit until the low-pressure switch shuts it off. Check the pressure in the system with a gauge. If the pressure does not go down near 0 pounds per square inch, set the low-pressure switch to a lower setting.

When the unit pressure is reduced to zero, shut the unit down. Remove the oil plug slowly, and observe the level of the oil in the crankcase through the hole. If the oil level is not up to the bottom of the filler hole, add oil to bring it up to the level recommended by the compressor's manufacturer.

Whenever the compressor is opened to check or add lubricating oil, make sure that it is only open for a minimum amount of time and that no dirt, contamination, or moisture gets inside. Even the air that gets in can carry water vapor. After oil has been added, or whenever the system has been opened, the part opened should be evacuated by a vacuum pump. Then, it should be recharged with new refrigerant before the unit is started up again.

Checking for Oil Acidity. When the oil level is observed, a sample of oil can be taken and checked for acidity. This check should always be made if contamination to the system is suspected from an overheated motor. Acid test kits, such as the kit pictured in Figure 15–4, are available in most refrigeration supply houses. Follow the instructions that come with the kit for checking the acid level of the oil. If the acidity of the oil is higher than it should be, follow the compressor manufacturer's recommendations for changing the oil and cleaning the system.

FIGURE 15–4
Acid test kit (Courtesy of Sporlan Valve Co.)

15.3

Condenser Section

A **condenser** is basically a heat exchanger in which the heat that is absorbed by the refrigerant during the evap-

Condenser

Air-cooled condensers

orating process is given off to the condensing medium. The heat given off in the condenser is equal to the heat picked up in the evaporating process plus the heat picked up through the compressor. As heat is given off by the condenser, the temperature of the refrigerant in the condenser drops below the saturation temperature. Because of the drop in temperature, the refrigerant vapor condenses to a liquid—hence the name *condenser*.

Air-cooled condensers are the only types covered in this text. The most common type of air-cooled condensers have forced-air, tube-and-fin construction. They have a copper or aluminum coil with aluminum fins on the outside of the tubing. A fan or a blower moves outside air across the coil. The aluminum fins on the coil increase the surface area exposed to the air.

Service problems that originate in the condensing section of an air conditioning or a heat pump system can often be identified by either higher-than-normal or lower-than-normal pressure in the high side of the system. These two conditions may be related to anything from the amount of refrigerant in the system to problems in the low-pressure side of the system. In this section, we limit our discussion to the two pressure problems in the high side of the system.

High-Pressure Problems in a System

A higher pressure than normal in the high-pressure side of a system can be an indication of problems. A common source of high pressure is an overcharge of refrigerant. An overcharge is indicated if the pressures on both the high side and the low side are proportionately high. If the high pressure is only found on the high-pressure side of the system, not on the low-pressure side, some other problem is indicated.

Chapter 13 discussed the proper way to check a system for an overcharge of refrigerant. Other service problems related to high pressure in an air conditioning or a heat pump system may be due to a high temperature of the air entering the system at the condenser coil, a low air quantity across the coil, a dirty or obstructed condenser coil, or noncondensable gases in the condenser coil. Each problem is discussed in the following subsections.

High-Temperature Condenser Air.   High-temperature inlet air on a condenser coil can result when the air leaving the coil short-circuits back into the inlet side. Short circuiting happens when the condenser is located in a closed space or when the condenser's air intake is too close to a vertical wall or other obstruction. A visual inspection of the location of the condensing unit will normally identify such a problem. But checking the inlet and the outlet temperature of the condenser air should verify it.

To check a unit to see whether high air temperature is causing a problem, measure the inlet and the outlet air temperature of the condenser with a pair of accurate thermometers while the unit is operating. The temperature of the inlet air should be the normal, outdoor air temperature. The temperature rise across the coil should be about 20 degrees. If the inlet temperature is higher than the outdoor temperature, then short circuiting should be checked. If necessary, the unit should be relocated to prevent recycling of condenser air. If the inlet temperature is normal, but the temperature rise is too high, the problem may be due to inadequate air flow across the coil.

Inadequate Airflow across the Condenser Coil. Inadequate airflow across a condenser coil reduces the rate of heat transfer from the coil to the condenser air. With reduced heat transfer, the condenser coil has more vapor and less liquid refrigerant in it than it should have. The added vapor causes the pressure to be higher than it would be normally. The reduction of air across the coil can be caused by a slipping condenser blower belt, a dirty condenser coil, or an obstruction of the coil.

Slippage in the Condenser Belt.   To check the condenser blower belt for slippage, observe the belt and pulleys as the condenser blower starts and while it is running. If slippage is occurring, it can normally be seen when the blower motor starts up. Slippage on start-up is often accompanied by a squealing noise.

If slippage is occurring, the unit should be shut off at the unit disconnect. Adjust the belt tension by moving the motor, in its mount, to increase the distance between the motor pulley and the blower pulley. On belt-drive

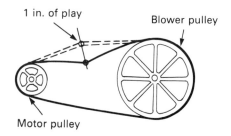

1 in. of play

Blower pulley

Motor pulley

FIGURE 15–5

Adjusting belt tension on belt-drive blower

blowers, the motor mounts are constructed so that the motor can be moved back and forth for belt adjustment. The belt should be adjusted so that it has about 1 inch of play between the two sides of the belt at a point midway between the two pulleys, as indicated in Figure 15–5.

Dirty or Obstructed Condenser Coils.   Condenser coils can get dirty from an accumulation of dust and dirt from the air that is drawn through them during normal operation. As dirt builds up on the outside of the coil surface, it insulates the refrigerant in the coil from the air on the outside of the coil. The dirt must be cleaned off to keep the coil operating efficiently.

A visual inspection will show if a condenser coil is dirty. If there is any doubt, take a sample by scraping across the coil tubing with a knife blade. If an accumulation of dirt is present, the coil should be cleaned.

A dirty coil can be cleaned with compressed air or a stream of water. If the accumulation is especially bad, chemical cleaning may be necessary. Special cleaning kits are available for cleaning the coils. These kits include the necessary chemicals for cleaning the coil and have complete instructions on the procedure to be used.

Blockage of the inlet side of a condenser coil can be caused by an accumulation of leaves, paper, or any trash that is blown by the wind or that is drawn against the coil by the inlet air. A visual inspection will usually show such an accumulation, and it can then be removed.

Noncondensable Gases in the Condenser Coils.   After the checks for slippage and dirt have been made, the unit may still have an abnormally high head pressure. If it does, then the condenser coil may have air in it. If air gets into the system through a leak in a refrigerant line or through the service work being done, it will normally accumulate in the condenser coil. Air remains in the gaseous state at the pressures found in air conditioning or heat pump systems, and it will not condense in the condenser coil.

Because air in the gaseous state takes up more space than refrigerant in the liquid state, it will cause an increase in pressure in the condenser coil and in the entire high-pressure side of the system. The amount of increase in the high-side pressure depends on the amount of air in the coil. The increase can amount to as much as 40 to 50 pounds per square inch above normal pressure.

To remove air from an air conditioning or a heat pump system, vent it from the condenser coil. If you cannot remove the air by venting, remove the refrigerant from the system. Then, evacuate the system with a vacuum pump.

Some large systems have an **air purge valve** on one of the top return bends of the condenser coil or on the liquid receiver. If the system being serviced has such a valve, it can be opened while the unit is operating, and the air can be purged. The valve should be opened carefully for short periods of time, and the air leaving it observed. When liquid refrigerant starts to come out of the valve, all of the air has been purged. The valve should then be closed tightly for normal operation of the system. If the system does not have a purge valve, it may have refrigerant line fittings at the top of the condensing coil. Loosen a fitting enough to vent the coil through the fitting.

To remove air from a small system that does not have a purge valve or line fittings, you must remove the refrigerant from the system. Then, evacuate the system with a vacuum pump to remove the air. If the system is small, the refrigerant can be voided to the atmosphere.

If the system is large, and you wish to save the refrigerant, you can pump it into a clean, dry, empty refrigerant cylinder. The refrigerant in the cylinder should then be cooled below the condensing temperature for the refrigerant, which causes the refrigerant to condense to a liquid. The air, being noncondensable, remains in the gaseous state in the top of the cylinder. The air can be vented from the cylinder by opening the valve on the top and allowing it to escape. When gaseous refrigerant starts to flow from the valve, the valve should be closed. The refrigerant will now be free of air. But before it is returned to the system, the entire system should be checked for leaks. If any leaks are found, they should be repaired. The system should then be evacuated thoroughly before refrigerant is replaced.

Air purge valve

Low-Pressure Problems in a System

Most problems due to low pressure in the high-pressure side of an air conditioning or a heat pump system are caused by either a low refrigerant charge or a low air temperature across the condenser coil. Chapter 13 covered the method of checking a unit for proper refrigerant charge. If during the check outlined there, you determine that the system is undercharged, then add refrigerant until the proper charge is reached.

If the temperature of the air entering the condenser coil is too low, the increased transfer of heat from the refrigerant in the coil to the air will cause the refrigerant to condense into a liquid too soon. Since the liquid takes up less space in the coil than the vapor does, the pressure in the coil will be lower than normal. This condition arises because of the placement of the condensing unit for an air-cooled system, which usually is located outside. Outdoors, the air is normally fairly warm during the cooling season. But if an air conditioning unit is operated during cool weather, the air across the condenser coil may be too cold for proper operation of the system.

The temperature of the air across the coil can be checked by measuring it with a thermometer. Compare the reading with the condensing temperature of the refrigerant being used. There should be about a 20-degree difference between the air and the condensing temperature of the refrigerant. If the air temperature is too low for proper operation, a low ambient kit can be used on the system. This kit includes a pressure switch controlled by the refrigerant pressure. The switch cycles the condenser blower off or on to maintain proper operating pressures.

15.4

Refrigerant Lines

There are two refrigerant lines in the high-pressure side of an air conditioning system: the hot gas discharge line and the liquid line. Service problems directly related to these lines are usually limited to kinks in the lines, leaks in the lines themselves, or leaks in the fittings used to attach the lines to the various parts of the system.

Kinks in the Lines

If a refrigerant line is bent too sharply around a corner, or if it is bent without the use of a mechanical bender, the line may kink. A **kink** is a point in the line where   **Kink** two sides of the tubing collapse toward each other. The cross-sectional area of a refrigerant line is reduced if the line is kinked. Refrigerant flow is reduced because of the extra friction produced by the restriction.

Visual inspection of the lines will usually show a kink, if one is present. To correct a kink in a line, replace the section of line with the kink in it. When replacing a line, observe all of the usual precautions related to pumping down and charging a unit.

Leaks in the Lines

If a refrigerant line is installed so that it touches another line or any metal part of the system, vibration can quickly rub a hole in the line. In such a case, refrigerant will leak through the hole.

If a hole is found in a refrigerant line, the refrigerant has to be taken out of that part of the system in order to reduce the pressure. Then, if the hole is not too large, it can be patched with silver solder or epoxy. If the hole is too large to be patched, the line should be replaced.

Leaks in the Fittings

Various methods are used to attach the refrigerant lines to the parts of a system. The most permanent method of connection is silver soldering. In this method, solder that contains silver is used. The parts to be joined are heated to a temperature high enough so that the solder will melt and join the parts. Other types of connections are also used, such as flare fittings and various types of quick-connects.

All types of fittings are subject to leaks. If a leak is suspected, the joints should be checked. There are various methods that can be used for checking for leaks. The three most common methods are the use of soapsuds, the use of a halide torch, and the use of an electronic leak detector. Checking a system for leaks by any of these three methods requires the unit to have some refrigerant in it so that there is pressure in the lines.

To check for leaks with soapsuds, use a concentrated solution of soap and water. Brush this solution on each joint, in turn. A leak will cause bubbles to appear in the soapsuds.

To look for leaks with a halide torch, light the torch. Move the sampling hose over the joints in the piping system. If any refrigerant is picked up by the sampling hose, the flame of the torch will begin to burn with a bright green color.

If an electronic leak detector is used, follow the instructions that come with the detector for setting it up for operation. Move the sampling tube over the fittings in the system, as illustrated in Figure 15–6. If any refrigerant is detected by the instrument, either a visual or

FIGURE 15–6
Technician using electronic leak detector (Courtesy of General Electric Co., Instrument Products Operation)

an audible signal will be given by the detector, depending on the type of detector used.

If any fittings are leaking, the refrigerant must be removed from that part of the system. The fittings should then be resoldered or replaced. After repairing the leaking fitting, charge the system with refrigerant. Check the joint again before starting the unit.

Service problems relating to the refrigerant lines themselves are relatively rare. General problems that may occur are limited to leaks, as covered in this section.

15.5

Accessories

Depending on the type of air conditioning or heat pump system being checked, there may be some accessories located in the refrigerant lines in the system. The most common accessories are sight glasses and liquid receivers.

The most common fault found with any accessory is the possibility of leaks in the fittings where the refrigerant tubing attaches to the accessory. Leaks in these joints should be identified and repaired in the same way as leaks in the piping fittings themselves are.

15.6

Summary

A service technician must know how to check the main components and parts of the high-pressure side of an air conditioning or a heat pump system in case of problems in the system. The proper operation of each of the parts is necessary for the proper operation of the entire system.

The main components of the high-pressure side of the system are the compressor, the condensing unit, and the refrigeration lines. In addition, there are various accessories to the parts.

Failure of the compressor, regardless of the type of compressor involved, is usually related either to mechanical failure of the compressor parts or to broken compressor valves. Mechanical failure is normally indicated when the compressor will not run. Valve failure is indicated when the compressor runs but does not raise the pressure in the system.

Service problems in the condensing section are indicated most often by higher-than-normal pressures in the system and sometimes, but rarely, by lower-than-normal pressures. Such problems are caused by a high or a low temperature of the condenser air or by inadequate air flow across the condenser coil. Before checking the components and parts in the high-pressure side of a system, a service technician should check the refrigerant pressure in the two sides of the system. This check determines whether the system has a proper charge of refrigerant.

Refrigerant lines and the accessories in the lines are relatively trouble-free. The most common faults found in lines and accessories are refrigerant leaks in the connecting fittings.

15.7

Questions

1. Open or semihermetic compressors can be dismantled for service, whereas a _____ compressor cannot.
2. Is the following statement true or false? A technician should mark the various parts of a compressor during dismantling so that each one can be replaced in exactly the same position it was in originally.
3. Is the following statement true or false? All crankshafts are the same in reciprocating compressors.
4. Is the following statement true or false? A compressor must be dismantled for a check of broken valves.
5. Is the following statement true or false? Any good-quality of motor oil can be used in a compressor for lubrication.
6. Lubrication of a compressor is provided by an ___ and by the oil _____ on the walls of the cylinders.

7. What causes the lubrication oil in a compressor to become acidic?

8. Service problems in the condensing section of a refrigeration system are often indicated by _____ pressure.

9. A common cause of high pressure in a refrigeration system is an _____ of refrigerant.

10. A blocked condenser coil or a lack of air across the coil will almost always result in _____ (high, low) pressure in the condensing section.

11. How do noncondensable gases, such as air in the condenser, cause high pressure in the system?

12. What are the two most common causes of low pressure in the condensing section of a system?

16

Expansion Device and Low-Pressure Side of the System

16.1

Introduction

As the part of the air conditioning or the heat pump system in which cooling or heating actually takes place, the low-pressure side of the system is extremely important to the operation of the entire system. In this text, the expansion device is considered as part of the low-pressure side of the system. The expansion device produces the pressure and the temperature drop in the refrigerant that allows it to boil off, or vaporize, in the evaporator coil. As the refrigerant vaporizes, it picks up heat from the air that is blown across the coil. A malfunction of the expansion device or any of the components of the low-pressure side of the system affects the operation of the entire system.

The major components of the low-pressure side of an air conditioning or a heat pump system are the expansion device, the evaporator assembly, and the refrigerant lines that connect these components to each other and to the other parts of the system. In addition, there are certain accessories whose functions are related to these components. These accessories are the distributor, the filter drier, and the suction line accumulator. This chapter describes these major components and accessories.

16.2

Expansion Devices

The **expansion device** in an air conditioning system is located in the liquid line just ahead of the evaporator coil. A heat pump system may have one expansion device that works with the refrigerant going in either direction, or it may have two expansion devices, one on each end of the liquid line just ahead of the coils. An expansion device produces a pressure drop in the refrigerant at the point at which it is located in the system. The pressure drop is accompanied by a temperature drop. This temperature drop cools the refrigerant so that it can pick up heat from the air that goes across the evaporator coil. This transfer of heat cools the air. The proper operation of the expansion device is of prime importance to the operation of the system.

Expansion devices may produce a pressure drop in one of three ways. The first way is by controlling the volume of the refrigerant flowing through the system. The second way is by controlling the pressure in the evaporator coil. The third way is by controlling the temperature of the refrigerant in the evaporator coil. Devices for each of these methods of control are described in the following subsections.

Expansion device

Volume Control Devices

Capillary tubes, float valves, and hand valves are the most common types of expansion devices used to control the volume of refrigerant flowing into the evaporator. Capillary tubes are used on small air conditioning or heat pump units, especially self-contained or packaged units. Float valves are usually found on flooded evaporator coil systems. Hand valves are used on large refrigeration and hydronic (circulating water) air conditioning or heat pump systems. These devices are discussed in the paragraphs that follow.

Capillary Tubes.   The capillary tube is the most common of the volume control devices used on small air

Capillary tubes

conditioning or heat pump systems. The capillary tube, as the name implies, is a small-diameter tube, but it is relatively long. The small diameter and the long length generate a considerable amount of resistance to the flow of refrigerant. This resistance creates a pressure drop between the liquid refrigerant ahead of the capillary tube and the liquid and vapor that leave the tube. The pressure drop causes some of the refrigerant to flash into a gas. This action reduces the temperature of the remaining liquid.

Since the capillary tube is a constant-volume control, it does not compensate for changes in the cooling load on the system. It is used most commonly on systems that do not have too much load variation. The size of the refrigerant charge in the system is critical when a capillary tube is used as the expansion device.

The pressures in the two sides of a system will equalize during an off cycle when a capillary tube is used as an expansion device. This equalization is an advantage. The compressor does not have to start against a high pressure, and so a simpler and less expensive compressor motor can be used.

Float valves

Flooded evaporator coil systems

Float Valves.   Many large refrigeration and hydronic air conditioning systems are **flooded evaporator coil systems**. That is, the evaporator is full of refrigerant in the liquid state during normal operation. Float valves are often used as expansion devices for these systems. The main purpose of the float valve is to maintain the liquid level in the system. The valve also provides a pressure drop between the liquid line and the liquid and vapor in the evaporator.

A float valve, illustrated in Figure 16–1, is a mechanical device. It has a needle valve, controlled by a float in the liquid, that opens or closes a port in the liquid line. If the liquid level rises, the valve closes and cuts off the flow of liquid. If the liquid level falls, the valve opens and allows more liquid to flow.

Hand valves

Hand Valves.   A hand valve, as the name implies, is an expansion valve that is adjusted by hand to regulate refrigerant flow. A hand valve used as an expansion device is a needle valve with very fine threads on the valve stem and body. The fine threads allow the user to

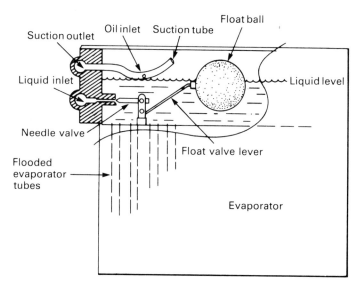

FIGURE 16–1
Float valve

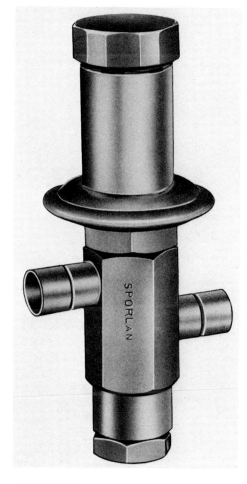

FIGURE 16–2
Automatic expansion valve (Courtesy
of Sporlan Valve Co.)

adjust the refrigerant flow very precisely. The needle in the valve adjusts against a seat. The valve is located in the liquid line of the system.

A system with a hand valve used as an expansion device requires constant supervision. The valve position is fixed wherever it is set. If the load on the system changes, adjustments to the valve setting must be made manually.

Hand valves, like float valves, are normally found only on large commercial or industrial applications. The need for an operator to be in attendance on the system rules out their use on small automatic systems.

Pressure Control Devices

Among pressure control devices, the most common device is the **automatic expansion valve**. Figure 16–2 shows a typical automatic expansion valve. The function of an automatic expansion valve is similar to the function of a pressure reducing valve. It is installed in the liquid refrigerant line, just ahead of the evaporator coil. It

Automatic expansion valve

maintains a constant pressure in the evaporator regardless of the cooling load.

An automatic expansion valve is used only on systems where the cooling load is not expected to vary. An automatic expansion valve will not compensate for changes in the load. This type of valve is often used on small refrigeration units. Because the valve does not compensate for varying loads, an automatic expansion valve will starve a coil for refrigerant if the system is heavily loaded. And it will flood the coil if the load is light.

Temperature Control Devices

Thermostatic expansion valve

A **thermostatic expansion valve** is the most commonly used device for controlling the temperature of the refrigerant in the evaporator coil. A typical thermostatic expansion valve is shown in Figure 16–3. It has a ball valve that is connected to a diaphragm. The valve is opened or closed as pressure on the diaphragm changes.

The diaphragm of the valve is connected to a remote bulb by a small hollow tube called a capillary tube, as shown in Figure 16–4. The bulb is attached to the suction line of the system. The bulb has a small charge of refrigerant in it. As the temperature of the refrigerant inside the suction line increases or decreases, the refrigerant in the bulb vaporizes or condenses. The change in refrigerant changes the pressure on the diaphragm. As the pressure on the diaphragm changes, the ball valve opens or closes. The action of the ball valve controls the flow of refrigerant into the coil.

A thermostatic expansion valve can be adjusted so that it will open or close at any given temperature sensed by the bulb, within its range of operation. It is usually set to maintain a certain amount of superheat—that is, a temperature rise above the boiling point—of the refrigerant used in the system.

A thermostatic expansion valve is an excellent expansion device for air conditioning or heat pump systems that have varying cooling loads. It will automatically adjust to the cooling load. That is, it will allow more refrigerant to flow if the load increases. And it will reduce the flow of refrigerant if the load decreases.

FIGURE 16–3
Thermostatic expansion valve (Courtesy of Sporlan Valve Co.)

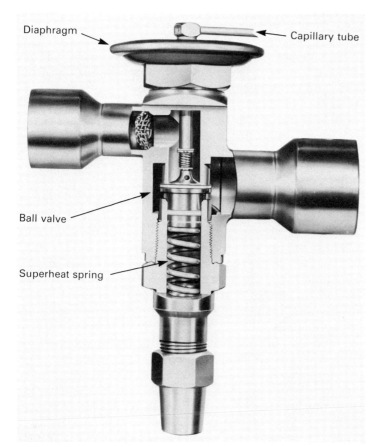

FIGURE 16–4
Cutaway view of thermostatic expansion valve (Courtesy of Sporlan Valve Co.)

FIGURE 16–5
Low-pressure side of air conditioning system, showing evaporator coil and blower

16.3

Evaporator Section

The **evaporator section** is the part of an air conditioning or a heat pump system in which the refrigerant vaporizes, or evaporates, as it picks up heat from the evaporating medium. The section consists of the evaporator coil, a blower to move air across the coil, and, in most cases, some accessories to other parts. A typical evaporator section is shown in Figure 16–5.

Evaporator section

Evaporator coils

Evaporator blower

Distributor line

Suction line

Refrigerant distributor

Most **evaporator coils** are formed of copper or aluminum tubing. The tubing has aluminum fins pressed onto it to increase its heat transfer efficiency. The refrigerant in the system flows through the inside of the coil, and air is blown across the outside by the evaporator blower. The coil is sized so that the proper amount of heat transfer will occur between the refrigerant and air, according to the cooling capacity of the unit.

When the system is operating, refrigerant enters the coil as it leaves the expansion device. The refrigerant enters the evaporator coil at the normal, low-side pressure. The temperature of the refrigerant is also lower. As the refrigerant goes through the coil, it picks up heat from the air going across the outside of the coil. This heat causes the refrigerant to vaporize.

The **evaporator blower**, shown in Figure 16–6, moves air across the evaporator coil. The blower is usually connected to the air distribution part of the system by a return duct on the inlet side and by a supply duct on the outlet side.

When a unit is connected to ductwork, a centrifugal blower is used. This type of blower moves the air most efficiently against the resistance in the duct system. If the air conditioning or heat pump unit has no ductwork, as is the case with a window unit, for instance, then a propeller fan may be used as the evaporator blower. This type of blower can move air efficiently when there is little resistance to airflow. The blower part of the evaporator section is covered more fully in Chapter 18.

The amount of cooling that an evaporator section will produce is a function of the design of the evaporator, the temperature and the quantity of refrigerant flowing through it, and the temperature and the quantity of air moving across the coil. A design engineer selects the proper evaporator section to fit the cooling needs of the application.

16.4

Refrigerant Lines

There are two refrigerant lines located in the low-pressure side of an air conditioning or a heat pump system. One line, the **distributor line**, runs from the expansion

FIGURE 16–6
Evaporator blower (Courtesy of Barry Blower Co.)

device to the evaporator coil. The other line, the **suction line**, runs from the coil to the compressor inlet. These two refrigerant lines are shown in Figure 16–7.

The distributor line from the expansion device to the coil is usually quite short, since it is desirable to have the expansion device located close to the coil. This line is sized to carry the refrigerant in the system, with about 20% to 30% of the refrigerant in the vapor state and the rest in the liquid state. The refrigerant that is in the vapor state is the part that boils off, or flashes into gas, as it goes through the expansion device.

The suction line runs from the evaporator coil to the compressor inlet port. It is larger than the other refrigerant lines in the system because it must carry the refrigerant in the vapor state. The refrigerant in the vapor state has a greater volume than it has in the liquid or partial liquid state.

If a thermostatic expansion valve with a remote bulb is used on a system, the bulb is attached to the suction line close to the evaporator coil. This bulb is attached to the outside of the line. It does not affect the flow of the refrigerant in the line at the point of connection.

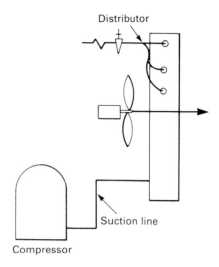

FIGURE 16–7
Refrigerant lines on low-pressure side

16.5

Accessories

Some of the typical accessories found in the low-pressure side of an air conditioning or heat pump system are the distributor, the filter drier, and the suction accumulator. These devices are discussed in the following subsections.

Distributors

A **refrigerant distributor** is used on large systems where refrigerant must be fed into the evaporator coil at more than one place. The distributor has one inlet for the refrigerant and many smaller outlets, as shown in Figure 16–8. A splitter inside the distributor divides the refrigerant stream into the smaller lines leading out of it. The

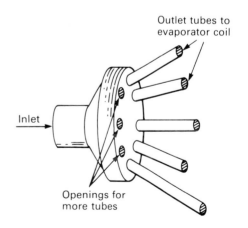

FIGURE 16–8
Refrigerant distributor

Filter drier

Suction accumulator

smaller lines feed the evaporator coil at several places. So, the refrigerant flows evenly throughout the evaporator.

Filter Driers

A **filter drier** is a combination filter and drier used in the liquid line ahead of the expansion device. It filters the refrigerant to remove any foreign particles that may be in it, and it dries the refrigerant to remove any water vapor that may be in it. Inside the filter drier is a fine-mesh screen filter on the inlet side. The rest of the drier is filled with a block of an absorbent chemical that traps and holds water vapor. Any refrigerant going through the liquid line must pass through the filter screen and then through the chemical block as it goes through the filter drier.

In small units, the filter drier may be an enlarged section of the tubing. Some units have a small canlike filter drier in the line, such as the filter drier shown in Figure 16–9.

In large units the filter drier is serviceable. For a serviceable filter drier, valves are installed in the refrigerant lines on each end of the filter. The valves can be used to isolate the filter from the rest of the system. If the filter needs cleaning, the valves can be closed, and the ends of the drier can then be removed. A new screen and a new chemical block can be installed in the drier without opening up the entire refrigeration system. A serviceable filter drier is often installed on a unit when a compressor is replaced because of a burnout. The system can be cleaned up by operating the unit and by using the filter drier to catch any dirty particles from the burnout that circulate through the system. The core of the drier is changed as often as necessary to clean out the system.

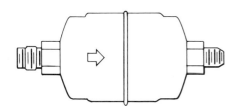

FIGURE 16–9
Filter drier

Suction Accumulators

A **suction accumulator** is a refrigerant reservoir in the suction line between the evaporator coil and the compressor. It is used to catch any liquid refrigerant that gets through the evaporator coil. Figure 16–10 illustrates a typical suction accumulator.

Liquid refrigerant must not be allowed to enter the compressor. If it does, hydraulic pressure will cause damage to the valves or other parts of the compressor. The suction accumulator stores any liquid that reaches it and allows only vapor to flow out.

The illustration in Figure 16–11 of the internal arrangement of an accumulator shows how the accumulator works. A suction accumulator is constructed so that any refrigerant that enters it in the liquid state is trapped in the accumulator. Only refrigerant in the vapor state flows out. The inlet line inside the accumulator is arranged so that the refrigerant is directed through a series of deflectors. The deflectors cause the vaporization of the liquid refrigerant.

Refrigerant can only leave the accumulator through an opening in the outlet line. This opening is located at the top of the tank. Any liquid refrigerant in the tank

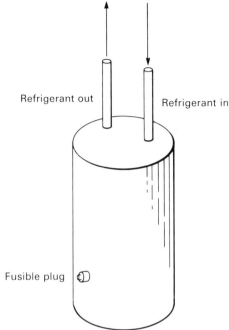

FIGURE 16–10
Suction accumulator

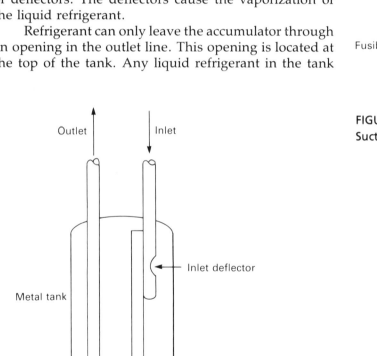

FIGURE 16–11
Cutaway view of suction accumulator

vaporizes before it reaches the level of the outlet. A small hole in the bottom of the outlet line, called a metering orifice, allows oil trapped in the accumulator to be drawn into the outlet. Most accumulators also have a fusible plug located on the side. The fusible plug will melt in case of a fire and allow the refrigerant to escape before dangerous pressure is built up.

16.6

Summary

The low-pressure side of an air conditioning or a heat pump system is the part of the system in which the refrigerant picks up heat from the medium to be cooled. The major sections of the low-pressure side of the system are the expansion device, the evaporator section, and the refrigerant lines. In addition, there are some accessories within the components that contribute to the proper functioning of the system.

There are several types of expansion devices that may be used in any particular system. Among these devices are capillary tubes, float valves, hand valves, automatic expansion valves, and thermostatic expansion valves. Expansion devices operate by controlling refrigerant volume, pressure, or temperature. While each of the different expansion valves functions in a slightly different way, each serves the same purpose in the system. That is, it causes a pressure drop in the system where the refrigerant enters the evaporator coil.

The evaporator section is made up primarily of the evaporator coil and the evaporator blower. The evaporator coil is the part of the evaporator section in which the transfer of heat from the cooling medium to the refrigerant takes place. The evaporator blower moves the air that is being cooled or heated across the coil. It is usually part of the air distribution system.

There are two refrigerant lines in the low-pressure side of the system. The distributor line runs from the expansion device to the evaporator. This line is usually very short, and it often contains a distributor to improve the efficiency of the evaporator coil. The second refrigerant line is the suction line, and it runs from the evaporator coil to the compressor suction valve. If a suction accumulator is used on a system, it is installed in the suction line.

In addition to the major sections and components, there are several accessories found in the low-pressure side of most air conditioning or heat pump systems. Among these accessories are distributors, filter driers, and suction accumulators.

Each section, component, and part of the low-pressure side of an air conditioning or a heat pump system is important to the operation of the system. If any one of these parts fails to function as it should, the entire system may shut down. Thus, an air conditioning and heat pump technician must understand the operation of each of these parts in order to troubleshoot or service them.

16.7

Questions

1. What is the main function of the expansion device in a system?
2. What are the three ways in which expansion devices produce a pressure drop?
3. Name three types of expansion devices that control refrigerant volume.
4. How does a capillary tube produce a pressure drop?
5. Does a capillary tube expansion device compensate for changing loads in the system?
6. Float valves are usually used on what type of systems?
7. Why are hand valves usually used only on large systems?
8. Name the most commonly used pressure control device used as an expansion device.
9. Why is an automatic expansion valve not used more often on typical air conditioning systems?
10. The most commonly used expansion device is a _____.
11. Describe a thermostatic expansion valve.
12. A thermostatic expansion valve controls _____ in the operating system.

13. The _____ of a system cools the air blown across the coil as the refrigerant in the system picks up heat from that air.
14. The heat picked up in the evaporator is given up to the _____.
15. Name three accessories commonly found in the refrigerant lines in the low-pressure side of a system.
16. Name the two functions served by a filter drier in a system.
17. Describe how a suction accumulator functions.

Servicing the Low-Pressure Side of the System

17.1

Introduction

A service technician should be able to check out each of the components and parts of the low-pressure side of an air conditioning or a heat pump system in case of a malfunction within the system. Understanding how each of these parts operate and how their operation affects the entire system is important. And knowing what steps to follow in the actual process of troubleshooting or servicing the parts is equally important. This chapter describes the steps for checking each of the components and parts of the low-pressure side of a typical air conditioning or heat pump system. Step-by-step procedures are explained for checking the expansion device, the evaporator section, the refrigerant lines, and the accessories found in the low-pressure side of the system.

17.2

Refrigerant Charge

Some operating problems that seem to originate in the low-pressure side of the system can be the result of either an overcharge or an undercharge of refrigerant. With a normally operating system, an **overcharge** is indicated by higher-than-normal pressure in both the low- and the high-pressure sides of the system. An **undercharge** is indicated by a lower-than-normal pressure in both sides.

Overcharge

Undercharge

Overcharge

To check for an overcharge, operate the system under conditions as nearly normal as possible. Check the pressures in the system with a set of gauges to see whether they correspond to those shown on the pressure/temperature operating chart furnished with the unit by the manufacturer. Figure 17–1 illustrates the connections for the manifold pressure gauge set. As shown in the figure, the compound gauge is attached to the suction service valve. The high-pressure gauge is attached to the discharge service valve.

If both the low- and the high-side pressures are high for the operating conditions, the unit is probably overcharged. If they are both low, the unit is probably undercharged.

To reduce the refrigerant charge in a system that has been overcharged, make sure that the compound gauge valve is closed, as indicated in Figure 17–2. Then, open the valve on the high-pressure side of the gauge manifold, and let some of the refrigerant flow out of the center hose.

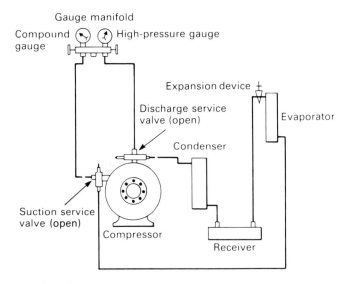

FIGURE 17–1

Attachment of manifold gauge set for measuring pressures in low side

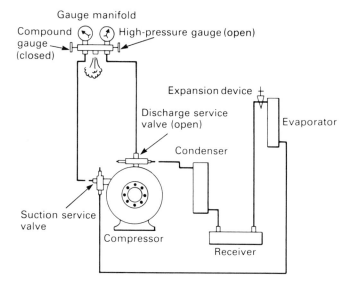

FIGURE 17–2
Valve positions for reducing refrigerant charge

If a large system is being serviced and a large amount of refrigerant is involved, the refrigerant can be saved. A special refrigerant cylinder that is designed for refilling must be used. Do not save the refrigerant unless you are sure that it is clean and dry.

To transfer the refrigerant from the unit to the cylinder, connect the service gauges to the unit and to the cylinder as shown in Figure 17–3. The pressure in the refrigeration system must be higher than the pressure in the cylinder. Start the unit. Open the valve on the high-pressure side of the service gauge manifold. Open the valve on the cylinder. Carefully watch the service gauges during the procedure so that you know when to stop. When enough refrigerant has been pumped out of the system so that the pressures in the two sides of the system are correct for the operating conditions, close the valves. Shut off the unit, and remove the gauge set.

Undercharge

To check for an undercharge in the system, attach the gauges as shown in Figure 17–1, and check the pressures

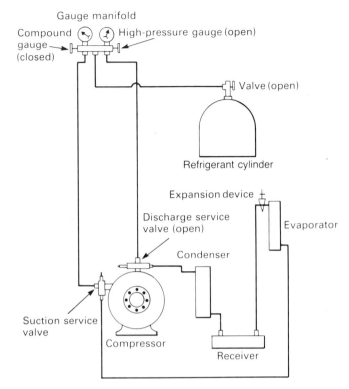

FIGURE 17–3
Valve positions for transferring refrigerant

while operating the system. If the pressures on both sides of the system are below those shown on the pressure/temperature operating chart for the operating conditions, the unit is probably undercharged.

Leaks in the piping system are the most common reason for a unit being undercharged. If a unit is found to be low on refrigerant, it should be checked for leaks. If any leaks are found, they should be repaired before the unit is charged.

If a leak is suspected in the low-pressure side of the system, make a visual inspection of the evaporator coils and all low-side refrigerant piping. A leak will often be indicated by the presence of oil. If refrigerant leaks out of the system, it carries with it a trace of compressor lubricating oil. Any oil stain on or around a refrigerant line or connection may be an indication of a leak.

If a leak is not found by a visual inspection, the system should then be checked with instruments. The procedures for leak testing were described in Chapter 13.

17.3

Expansion Devices

In modern air conditioning or heat pump practice, a wide variety of refrigerant control valves are used as expansion devices to obtain the most efficient operation of a system. Among these devices are capillary tubes, hand valves, float valves, automatic expansion valves, and thermostatic expansion valves. Service procedures for each of these devices are covered in the following subsections.

Capillary Tubes

A **capillary tube** has a small diameter but a relatively long length. When a capillary tube is used as an expansion device, it is subject to blockage because of the very small bore of the tube. Any small metal burr or particle of dirt in the refrigerant can block the tube. Such blocking particles get into the system when care is not taken to keep the refrigerant lines clean during installation or service procedures.

    A restricted capillary tube is indicated if the system does not produce the proper cooling effect. It may also be indicated by low refrigerant pressure in the low-pressure side of the system and by high pressure in the high-pressure side. One way to check the capillary tube for a restriction, without removing it from the system, is to observe the frost line on the tube while the unit is operating. The **frost line** is the point on the refrigerant line where frost first appears. A normally acting capillary tube has frost only on the leaving end of the tube. A restricted tube will have frost on the tube from the point of restriction to the end of the tube.

    Another way to check the operation of the capillary tube is to measure the temperature along the tube during

Capillary tube

Frost line

operation of the unit. There should be a gradual temperature drop along the length of a properly operating tube. There will be a pronounced drop in temperature at the point of restriction in a partially blocked tube.

A restricted capillary tube should be removed from the system. The best way to clear the obstruction is by applying air pressure. If the restriction cannot be cleared from the tube, then the tube should be replaced with a new one of the same bore diameter and length as the original.

Float Valves and Hand Valves

In some specialized applications, a volume control device is desirable for the operation of an air conditioning or a heat pump system. Float valves and hand valves are used as this type of control device. **Float valves** are mechanical valves in which the flow of refrigerant is controlled by the level of the liquid refrigerant in a tank. A **hand valve** is a needle valve that is operated by hand to control the refrigerant flow.

Float valves are normally mounted on a float chamber that is adjacent to, or part of, a flooded coil evaporator, as illustrated in Figure 17–4. To determine whether a float valve is operating correctly while the system is running, check the pressures and temperature of the refrigerant in the high and the low sides of the system. Use a pressure/temperature chart to determine whether they are correct for the system and for the application.

Since the float valve is a mechanical device, the best way to check it is to visually inspect it. But inspection can only be done by dismantling the valve. To dismantle it, shut the equipment down. Pump down the system's refrigerant into the condenser section. Close the service valves to isolate the float valve. Remove the float valve, dismantle it, and inspect the operating parts. Replace worn parts before reassembling the valve. Evacuate the low-pressure side of the system before recharging.

The operation of a hand valve is checked by the same methods used for a float valve. That is, one of two methods can be used. Either observe the pressures and

Float valves

Hand valve

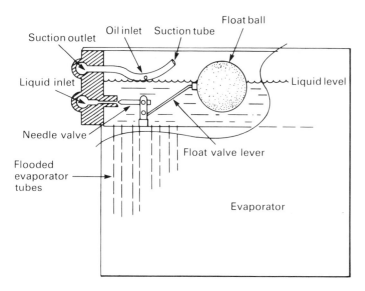

FIGURE 17–4
Float valve on flooded coil system

temperatures in the two sides of the system during operation. Or visually inspect the disassembled valve while the unit is shut down.

Automatic Expansion Valves

An **automatic expansion valve** controls the pressure in the low side of the system regardless of the cooling load or the condensing air temperature. It is usually used on systems where the cooling load is fairly stable and does not change significantly during operation.

    The best way to check the operation of an automatic expansion valve is to check the pressure on the low-pressure side of the system while the unit is operating. Since the valve is designed to maintain a constant pressure in the evaporator coil, the pressure in the low side, as measured with the service gauge, should be the same as the pressure indicated on the adjustment stem on the valve. To check this setting, remove the cap that covers the valve adjustment stem and observe the setting. It will be marked in pounds per square inch gauge (psig).

Automatic expansion valve

If the pressure in the system is not within 1 or 2 pounds of the indication on the stem—or of what is desirable in the system—try adjusting the setting by turning the stem in or out. If the pressure in the evaporator cannot be adjusted to match the valve setting, the valve should be replaced.

In some cases, the operation of an automatic expansion valve can be checked by observing the frost on the distributor lines entering and leaving the evaporator coil. In normal operation, there should be little or no frost on either line. Excessive frost on the distributor lines indicates a low pressure and a low temperature in the evaporator. Excessive frost on the suction line indicates high pressure and high temperature in the coil.

If excessive frost appears on either set of lines, the valve should be adjusted to produce balanced operation during a time when a normal cooling load is applied. Adjust the valve by turning the valve stem in or out with a valve wrench, and observe the frost line. The unit should be allowed to run for several minutes between each adjustment. In this way, you can make sure that the pressure in the evaporator relates to the operation of the valve. If the pressure in the low-pressure side of the system cannot be regulated by the valve, or if it will not hold steady after adjustment, the valve should be replaced with a new one.

Thermostatic Expansion Valve

Thermostatic expansion valve

The most commonly used device for controlling refrigerant flow into the evaporator is the thermostatic expansion valve. A **thermostatic expansion valve**, shown in Figure 17–5, is controlled by the temperature of the refrigerant leaving the evaporator coil. It is the most commonly used valve because it automatically adjusts for changing cooling loads or for changing temperatures of the condensing air.

A thermostatic expansion valve controls the operation of a system by maintaining a certain amount of superheat in the refrigerant as it goes through the evaporator coil. Procedures for adjusting the valve to obtain the correct superheat and procedures for other tests and servicing are described in the following subsections.

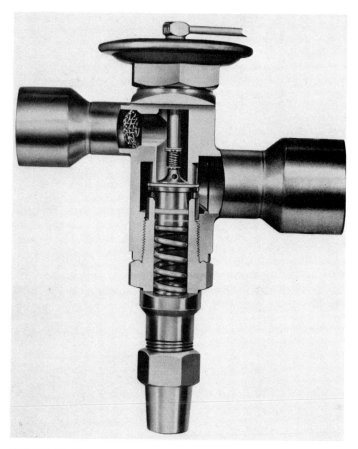

FIGURE 17–5
Cutaway view of thermostatic expansion valve (Courtesy of
Sporlan Valve Co.)

Adjusting for Superheat.   To check proper performance
of a thermostatic expansion valve, measure the pressure
on the low side of the system during normal operation.
Then, find the saturated temperature for this pressure
on a pressure/temperature chart for the refrigerant being
used. Next, measure the temperature of the refrigerant
leaving the evaporator coil. Compare the saturated temperature from the chart with the actual temperature measured. The difference between the two temperatures is
an indication of the **superheat** in the refrigerant as it   **Superheat**
leaves the evaporator coil. The superheat should run
from about 5 to 15 degrees Fahrenheit in most systems.

A thermostatic expansion valve can be adjusted for output of superheat by turning the adjusting stem of the valve in or out. If measurements show that the superheat is incorrect on a unit, adjust the valve until the proper superheat is obtained. Be sure and run the unit long enough for the operation to stabilize.

Pressures on the Valve.   Malfunction of a thermostatic expansion valve may cause a system to operate with either excessively low pressure in the low side or with excessively high pressure in the low side. The type of malfunction of the valve determines the pressure variation. The three types of pressure applied to a thermal expansion valve are indicated in Figure 17–6. They are pressure from the refrigerant in the remote bulb, pressure from the refrigerant in the evaporator coil, and superheat spring pressure.

If low suction pressure, as indicated by the low-side pressure gauge, is caused by the expansion valve,

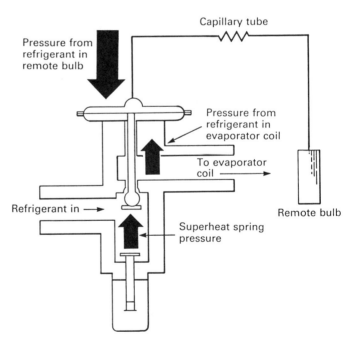

FIGURE 17–6

Pressures applied to thermostatic expansion valve

it may be an indication of either a dead power element or a restriction in the valve. To check the power element, remove the thermal-sensing remote bulb from the suction line. Warm the element by holding it in your hand. If the element is good, warming the bulb will cause the expansion valve to open and more refrigerant will flow through it. As a result, the low-side pressure will go up. If there is no change in the pressure during this test, the power head should be checked visually. For this check, the valve will have to be taken apart.

Dismantling the Valve and Testing the Power Head. To dismantle a thermostatic expansion valve, you must first pump down the system to lower the pressure in the low side. To pump down the system, close the service valve on the high-pressure side of the system by frontseating it. Operate the unit until it shuts off on low pressure. Jumper the low-pressure control, if necessary, to remove the refrigerant from the low-pressure side of the system. Shut the unit down. Make sure that the pressure has been reduced. Now, the expansion valve can be dismantled.

Figure 17–7 illustrates the parts of an expansion valve. Remove the bottom seal cap and bottom cap assembly from the valve. Remove the spring guide, spring, pushrods, and pin carrier. Check all of the removable parts for dirt and defects. Clean the valve body and all parts. Check the packing around the pushrods. Make sure that the pins move freely without allowing refrigerant to escape. The power head should now be exposed for checking.

Test the power head diaphragm by depressing it with your thumb. If the diaphragm cannot be moved, then the power head is good. Movement in the diaphragm during this test indicates a defective power element. If it is defective, it should be replaced.

Reverse the dismantling procedure to reassemble the expansion valve.

Checking for Restrictions in the Valve. Restrictions in a thermostatic valve are usually found to be due to either moisture freezing in the valve or a clogged filter screen in the valve. If there is any water vapor in the refrigerant in the system, it can freeze in the orifice of the expansion

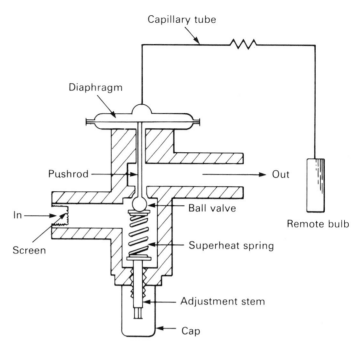

FIGURE 17–7
Parts of thermostatic expansion valve

valve while the refrigerant is flowing through the valve. Figure 17–8 shows where ice may accumulate in the orifice. This ice will block the flow of the refrigerant. The result is the same as what would occur if the needle valve in the expansion valve were to close. When the system is shut off, the ice melts and the blockage no longer exists. Thus, water vapor in the system may be suspected of causing erratic operation of the thermostatic expansion valve. In this case, the refrigerant should be removed from the system and the system evacuated by a vacuum pump before the valve is condemned.

A thermostatic expansion valve has a very fine-screen filter on the inlet side of the valve. The purpose of this filter is to prevent metal burrs from the tubing or dirt and contaminants from the refrigerant from entering the expansion valve. If there is much dirt in the system, it will eventually plug the screen and obstruct the flow of the refrigerant. If the screen is suspected of being dirty, the valve must be disassembled, as described previously. Then, the screen can be cleaned.

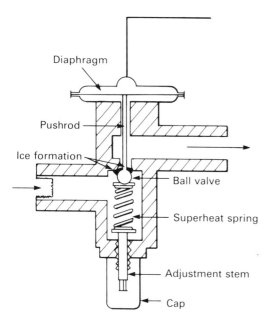

FIGURE 17–8
Ice formation in expansion valve

If the pressure gauge shows an abnormally high pressure on the low-pressure side of the system, the trouble may be an expansion valve that is sticking open because it is out of adjustment. This condition is often accompanied by a frosted or sweating suction line. The frost or sweat on the suction line shows that too much liquid refrigerant is flowing, and it cannot all evaporate in the coil. Some is passing through the coil and into the suction line beyond. The valve should be disassembled and inspected if it is suspected of sticking open.

17.4

Evaporator Section

The **evaporator section** is the part of the low-pressure side of the air conditioning or heat pump system in which the liquid refrigerant boils and evaporates. Here, the refrigerant picks up heat from the cooling medium as it changes into a vapor. Figure 17–9 shows an evap-

Evaporator section

FIGURE 17–9
Evaporator section of air conditioning or heat pump system
(Courtesy of Rheem Mfg. Co.)

orator section of an air conditioning or a heat pump system. The evaporator system includes the evaporator coil and the evaporator blower. This section covers the evaporator coil. The evaporator blower is covered in Chapter 18, along with other parts of the air distribution system.

Service problems directly related to the operation of the evaporator coil are not common. Physical damage to the coil in the form of punctures, pinching of the tubing, or leaks in the return bends are the most common source of problems. These problems can usually be located by physical examination or by the use of a leak detector.

In a normally operating system, the volume of air moving across the evaporator coil is fixed. With the proper refrigerant charge, the system produces a fixed refrigeration effect. This effect, in turn, causes a fixed temperature drop in the air going across the coil. If the refrigerant charge has been checked and is correct, measuring the temperature drop of the air across the coil is one way of determining how much air is moving across the coil. To check the temperature drop of the air, use two thermometers. Measure the temperature of the air entering the coil and of the air leaving. The difference between the two is the temperature drop. A normal drop is approximately 20 degrees. If the temperature drop is higher than normal, but the system has the correct refrigerant charge, then the air quantity is low. The airflow should be checked.

The most common visible result of insufficient airflow over the evaporator coil is frosting or icing of the coil. Frost on the coil is an accumulation of condensed moisture; frost is a snowy, crystallike covering on the outside of the coil. A covering of frost has an insulating effect, and it prevents the proper transfer of heat from the evaporator air to the refrigerant. If the frost does not melt, it will result in icing. Ice on the coil is a coat of frozen condensate. Frosting or icing of the evaporator coil causes a system to run with a very low pressure in the low-pressure side. This condition is damaging to the compressor and can cause compressor failure due to overheating. If frosting or icing occurs on an evaporator coil, the airflow across the evaporator coil should be checked.

Low airflow can be caused by slipping evaporator blower belts, dirty filters, dirty evaporator coils, or blockage in the duct system. Any of these defects can create problems in the evaporator section of the system.

Slipping evaporator blower belts can be checked visually and physically. Adjustments can be made as necessary to provide the proper tension. The belts should be adjusted so that there is approximately 1 inch of play between the two sides of the belts at a point midway between the two pulleys. This adjustment is described in detail in Chapter 18.

A dirty evaporator coil should be cleaned with compressed air or a water stream. If it is very dirty, it should be cleaned with chemicals.

Suction line

Distributor line

17.5

Refrigerant Lines

There are two sets of refrigerant lines on the low-pressure side of an air conditioning or a heat pump system. They are the line that goes from the expansion device to the evaporator coil and the **suction line** that goes from the evaporator coil to the compressor. In some systems, the line from the expansion device to the coil has a distributor in it and is called the **distributor line**.

Service to either of the refrigerant lines on the low-pressure side of the system is limited to a physical inspection or to a test for leaks in the lines or connections. Dents or kinks in the lines will cause a restriction in refrigerant flow. If a refrigerant line rubs against any other metal part of the system, a hole can quickly be worn in the tubing, causing a leak. Loose connections or poorly formed or welded joints in the tubing are also sources of leaks. If leaks are suspected, the piping and connections should be inspected.

17.6

Accessories

The three most common accessories in the low-pressure side of the system are the liquid distributor, the filter drier, and the suction accumulator. Service procedures for these accessories are described in the paragraphs that follow.

A **liquid distributor** is a source of trouble only if it gets plugged or if the connections to the liquid line or the evaporator coil develop leaks. The distribution lines from the distributor head to the coil are made of small-diameter tubing. If any contaminant in the refrigerant gets to one of these lines, there is a possibility that the line will get plugged. In this case, no refrigerant will flow to the part of the coil fed by that particular line.

To determine if a line is plugged, check the temperature of the line at several points between the distributor head and the coil. If the line is clear, the temperature drop along the line will be gradual. If the

Liquid distributor

line is blocked, there will be little or no temperature drop except at the point of the blockage. If any of the lines of the distributor are blocked, the distributor should be removed from the system. The lines should then be cleared with compressed air, if possible. If you cannot clear the lines of the blockages, then a new distributor should be installed in the system.

Figure 17–10 shows a cutaway view of a **filter drier** and its absorbent block. A filter drier is normally not a source of trouble in a system. However, contamination in the system may block the filter, or water vapor in the refrigerant may saturate the drier. If a filter drier needs to be checked for either blockage or saturation, the refrigerant in the system should be pumped down. Then, the filter drier should be removed from the line. In a small system, it is good economy to replace the filter drier with a new one rather than attempt to clean the old one. In large systems, the filter drier will probably have a replaceable core, and this core can be replaced if necessary.

A **suction accumulator** is used in an air conditioning or a heat pump system to prevent any liquid refrigerant that may get through the evaporator without vaporizing from entering the compressor. The accumulator traps and stores any liquid refrigerant that enters it while allowing the vapor to pass through. A cutaway view of an accumulator is shown in Figure 17–11. The accumulator is designed so that it will allow any oil that is in the refrigerant to pass on to the compressor. The accumulator is located in the suction line of the system between the evaporator coil and the compressor.

Normally, the only problems related to the accumulator are leaks that may occur in its refrigerant line connections. If any leaks are suspected, they should be checked in the usual way. The leaks should be repaired before the unit is charged.

17.7

Summary

A service technician should be able to check the individual sections, components, and parts of the low-pres-

Filter drier

Suction accumulator

FIGURE 17–10
Cutaway view of filter drier (Courtesy of Superior Valve Co.)

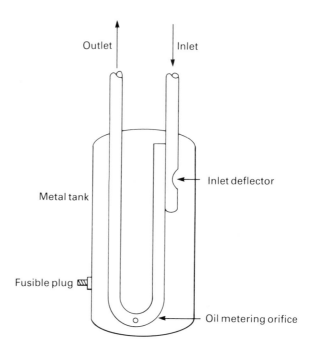

FIGURE 17–11
Cutaway view of suction accumulator

sure side of an air conditioning or a heat pump system in case of a malfunction in the system. Different service procedures are used for checking each section or component.

The refrigerant charge is related to the entire system. But some of the effects of an undercharge or an overcharge of refrigerant may be seen primarily on one side of the system. A low refrigerant charge is identified by low pressure, and it is often caused by leaks in the system. An overcharged system is usually related to high pressure in the system.

Proper operation of an air conditioning or a heat pump system depends on the proper operation of the expansion devices. There are several types of expansion devices used. The most common devices are capillary tubes, float valves, hand valves, automatic expansion valves, and thermostatic expansion valves. The proper method for checking each of these devices is usually a combination of a visual inspection and an observation of the pressure and temperature in the low-pressure side of the system.

The evaporator section in an air conditioning or a heat pump system includes the evaporator coil and the evaporator blower. Service problems in the coil are usually limited to leaks or physical damage to the tubing in the coil or to leaks in the connections to the coil. Service of the blower section of the evaporator is covered in Chapter 18.

There are two refrigerant lines in the low-pressure side of a system. They are the line from the expansion device to the coil and the suction line from the coil to the suction side of the compressor. The main service problems related to the refrigerant lines are physical damage or leaks.

Most air conditioning or heat pump systems have some accessories in the low-pressure side of the system. Among these accessories are liquid distributors, filter driers, and suction accumulators. Contamination in the system and leaks in the refrigerant line connections to these devices are the most common problems related to them.

17.8

Questions

1. To check an air conditioning or a heat pump system to see whether it has the proper amount of refrigerant, you connect a _____ to the unit and read the pressures.
2. Is the following statement true or false? To check the pressures in a system, you must connect a cylinder of refrigerant to the system.
3. What is the most common cause of an undercharge in a system?
4. Why will there often be an oil stain at the point of a leak in a refrigerant line?
5. One way of visually checking a capillary tube for a restriction is to look for the _____ on the tube during normal operation.
6. An automatic expansion valve works by regulating the _____ of the refrigerant in the low-pressure side of the system.

7. Describe what is meant by *superheat*.

8. Is the following statement true or false? A thermostatic expansion valve is nonadjustable.

9. Is the following statement true or false? A thermostatic expansion valve must be taken apart in order to check the operation of the power head diaphragm.

10. Suppose a system is properly charged with refrigerant. In this case, what do you determine when you measure the temperature drop across the evaporator coil?

11. Name three causes of low airflow across an evaporator coil.

12. How much play should there be between the two sides of a properly adjusted blower belt?

Air Distribution System

18.1

Introduction

An **air distribution system** for an air conditioner or a heat pump includes the evaporator blower, the supply and return air ducts, the supply registers, the return air grilles, and any dampers or other air control devices installed in the ductwork. To perform properly, the air distribution system must be correctly designed, installed, and balanced. The final performance of the entire air conditioning or heat pump system is dependent as much on the air distribution system as it is on any other part of the system. Poor performance of the air distribution system can cause poor operation of the entire system.

Design of the air distribution system is actually the job of the systems design engineer. But proper air distribution is so important to the operation of the total system that a service technician should be able to check the air system design as well as the mechanical parts of a system.

The major parts of the distribution system, the function of the parts, and the proper procedures for troubleshooting or servicing each of the parts are described and explained in this chapter.

Air distribution system

18.2

Blower

One of the most important parts of the distribution system is the blower, or the air mover. The **blower** moves

Blower

Belt-drive blower

Direct-drive blower

Cubic feet per minute, cfm

the air that is being cooled or heated through the evaporator or the indoor coil. Then, it circulates the same air through the building. In most small systems, the system blower is also the evaporator blower. But in large systems, the blower may be separate from the evaporator, although it is located in the same duct system.

The air conditioning or heat pump system blower is usually a forward-curved, centrifugal blower powered by an electric motor. A forward-curved blower is one in which the blades are inclined in the direction of the airflow. Figure 18–1 illustrates a typical evaporator blower.

In small packaged units, a direct-drive blower is usually used. In most large units, a belt-drive blower is used. In a **direct-drive blower**, the blower wheel is connected directly to the motor shaft. In a **belt-drive blower**, the motor is mounted next to the blower wheel but is separate from it. The motor and the blower are connected by pulleys and belts.

The following subsections discuss air delivery, service procedures related to air delivery, and adjustments for the blower.

Air Delivery

Each air conditioning or heat pump unit is sized to provide a certain amount of cooling or heating. The evaporator or indoor coil is sized to produce that effect with a given amount of air moving across it. The airflow is measured in **cubic feet per minute**, or **cfm**. This measurement describes the air that is circulated through the building. The blower in the system is selected so that it will move the air required against the resistance to the flow found in the system. If the quantity of air produced by the blower drops below the amount required for proper operation of the unit, the operation of the entire system is affected.

Two things can affect the cfm output of a blower. The first is a reduction in the speed of rotation of the blower wheel. The second is a restriction in any part of the air system. Both effects are discussed in the following subsection.

FIGURE 18–1
Evaporator blower (Courtesy of The Trane Co.)

Effects on cfm Output.   A reduction in the speed of the blower wheel is usually caused either by incorrect wiring connections on a direct-drive blower or by blower belt slippage in belt-drive systems. A complete lack of airflow is usually caused by an interruption to the electric power supply, a blower motor failure, or a blower belt breakage.

Restrictions to the airflow can occur in any part of the air distribution system. If we assume that the duct system is properly designed and installed, then the place to start checking the system for restrictions is at the evaporator or indoor coil. An obvious symptom of low evaporator airflow is ice on the evaporator or indoor coil. The ice may be a slick, smooth covering on the surface of the coil. It may be a very thin coat on the coil surfaces, or it may build up until it becomes a solid block. Ice may cover the coil surface only, or it may build up over the liquid line and the distributor. It may even extend to the suction line leaving the coil. If the suction line ices up, it is a symptom of a problem that can damage the compressor. Ice on the suction line indicates that there is liquid refrigerant in the line, and this condition can cause liquid slugging of the compressor.

There are two ways to check the quantity of airflow across the evaporator coil to see whether airflow is restricted. One way is to measure the pressure drop in the air across the coil and to use a pressure drop chart furnished by the manufacturer to determine the cfm. The second way is to measure the temperature drop of the air across the coil and to calculate the cfm by a formula. Each of these methods is described in the subsections that follow.

Measuring Pressure Drop.   A **pressure drop chart** is furnished with each new air conditioning or heat pump unit by the manufacturer. This chart shows the pressure drop across the coil for a wide range of cfm. The charts are developed from laboratory data for actual performance. Typical pressure drop charts for a hydronic coil are shown in Figure 18–2. The top chart gives the pressure drop, and the bottom chart gives the correction factors for a specific model.

To measure airflow across a coil by the pressure drop method, use a manometer to read the static pres-

Pressure drop chart

Accessory Heating Coil Airside Pressure Drop* for Fan-Coil Units (Models BU & BH)

CFM COMPARED TO RATED QUALITY		−20%	−10%	STD.	+10%	+20%
BU70	STEAM	0.165	0.200	0.237	0.278	0.326
	WATER	0.165	0.200	0.237	0.278	0.326
BH100	STEAM	0.300	0.360	0.43	0.52	0.60
	WATER	0.220	0.255	0.32	0.38	0.44
BH150	STEAM	0.37	0.46	0.55	0.64	0.75
	WATER	0.28	0.35	0.40	0.48	0.55

*Inches of water.

Heating Coil Correction Factors for BU 70*

STEAM	0 LBS.		2 LBS.		5 LBS.		10 LBS.**	
	0.96		1.0		1.05		1.12	
HOT WATER			ENTERING WATER TEMP.					
	TEMP. DROP	160°		180°		200°		220°
	10°	0.76		0.96		1.20		1.40
	20°	0.60		0.80		1.00		1.20
	30°	0.45		0.62		0.88		1.02

* For BH100 and BH150 — see Trane Heating Coil Catalog.
** Maximum recommended pressure.

FIGURE 18–2   Pressure drop charts for typical coil (Courtesy of The Trane Co.)

sure on each side of the coil. To use a manometer, drill or punch holes in the ductwork upstream and downstream from the coil. The holes should be in the center of the side of the duct. Make sure that the holes go through the insulation in the duct. Level the manometer, and attach hoses to both ends of the U tube. Use pressure taps on the ends of the hoses, as shown in Figure 18–3. Place the taps in the holes. Make a good seal around the holes by using permagum around the taps where they fit into the holes. Make sure that the hose from the lower side of the manometer goes into the hole on the downstream side of the coil. With the blower running, read the static pressure difference on the manometer. The pressure indicated on the manometer gauge is the pressure drop across the coil.

Now, find the cfm corresponding to the pressure drop by checking the pressure drop chart for the unit. If the cfm you find is not the cfm needed, then the motor speed should be changed if it is a direct-drive blower. The pulleys sizes should be changed if it is a belt-drive blower. These changes allow you to obtain the correct

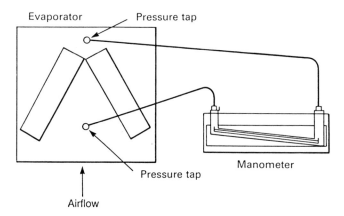

FIGURE 18–3
Measuring pressure drop across evaporator coil

cfm for the system. The manometer can be left attached while the drive changes are made. The cfm should be checked again after the changes.

Measuring Temperature Drop.   The second way to check air quantity across the evaporator coil is by the temperature drop method. In this method, the temperature drop across the coil is measured by thermometer readings, and the cfm is calculated from a formula. To do this check accurately, you must know the cooling or heating output, in British thermal units per hour (Btu/h), of the unit. The output rating is found on the manufacturer's data sheet for the equipment.

Measure the inlet and outlet air temperatures of the coil while the unit is running. Figure 18–4 shows the positions of the two thermometers for this measurement. The difference between the two readings is the temperature drop. The temperature difference should be approximately 20 degrees. For cooling, if the drop is much more than 20 degrees, then less air than normal is moving over the coil. If the temperature difference is less than 20 degrees, then more air than there should be is moving over the coil.

The actual cfm of air moving across the coil can be determined by using a formula based on the sensible heat in the air. If the cooling capacity of the unit is

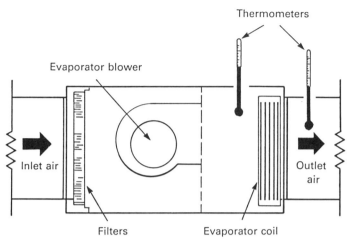

Thermometers

Evaporator blower

Inlet air

Outlet air

Filters   Evaporator coil

FIGURE 18–4
Measuring temperature drop across evaporator coil

known, and if the latent heat capacity is assumed to be approximately 30% of the total, then the following formula can be used.

$$cfm = \frac{Btu/h \times 0.70}{1.08 \times t_d}$$

where

cfm = air quantity across the coil, in cubic feet per minute

Btu/h = cooling capacity of the unit, in British thermal units per hour

0.70 = sensible heat capacity

1.08 = a constant, in British thermal units per cubic feet per minute per degree Fahrenheit

t_d = temperature drop across the coil, in degrees Fahrenheit

The following example illustrates the use of this formula.

EXAMPLE

A unit is rated at 60,000 British thermal units per hour (5 tons) of cooling. The air temperature drop across the coil is measured at 18.4 degrees Fahrenheit. Use the formula to find the cfm across the coil.

Solution

Use the formula.

$$\text{cfm} = \frac{60,000 \times 0.70}{1.08 \times 18.4} = 2113.5 \text{ cfm}$$

When the cfm through the coil is checked, it may not be the correct cfm for the capacity of the unit. Then, adjustments should be made to the blower to obtain the correct cfm.

Adjusting the Blower

The characteristics of a centrifugal blower—the type used in most air conditioning or heat pump units—are such that the quantity of air moved, in cfm, is directly proportional to the **revolutions per minute (rpm)** of the blower wheel. For an increase in the volume of air being delivered, the speed of rotation of the wheel is increased. For a decrease in the volume of air, the speed is decreased. Since the cfm of the air is directly related to the rpm of the blower, the relationship between these two characteristics can be given mathematically, as follows:

Revolutions per minute, rpm

$$\frac{\text{cfm}_1}{\text{cfm}_2} = \frac{\text{rpm}_1}{\text{rpm}_2}$$

where

cfm_1 = original blower output
cfm_2 = new blower output
rpm_1 = original blower speed
rpm_2 = new blower speed

If a direct-drive blower is used—that is, one with the wheel mounted directly on the motor shaft—the blower speed is the same as the motor speed. Most of the motors used for this type of application are multi-speed motors. There is a different electric connection to

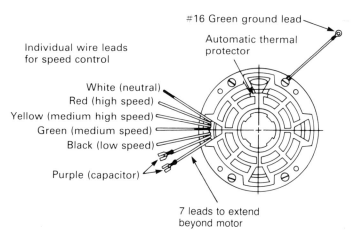

FIGURE 18-5
Direct-drive motor with speed control leads marked

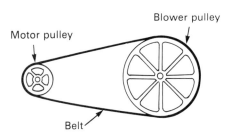

FIGURE 18-6
Belt and pulleys for belt-drive blower

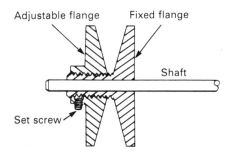

FIGURE 18-7
Adjustable pulley

the motor for each speed desired. To change the motor speed, you connect the electric power leads to different terminals on the motor. For convenience, the terminals for each speed are marked on the motor terminal board. Figure 18-5 shows a typical motor terminal board with the different terminals marked. The speeds are constant for each terminal connection. So, the closest speed to the one that will give the desired cfm from the blower must be used.

To change the cfm on a unit with a belt-drive blower, you change the rpm of the blower by changing pulley sizes. The blower pulley is usually two to three times the diameter of the motor pulley, as indicated in Figure 18-6. Because of the size difference in the pulleys, the blower turns more slowly than the motor. When you must change the speed of the blower, you change the size of the motor pulley.

Sometimes, adjustable motor pulleys are used. In this case, the speed of the blower can be changed by changing the diameter of the adjustable pulley. To change the diameter, loosen the set screw on the pulley. See Figure 18-7. Turn the outer part of the pulley in or out on the threaded hub. Turning the outer part of the pulley clockwise moves it in closer to the inner part. Turning the outer part counterclockwise moves it out away from the inner part. Moving the outer part of the pulley changes the effective diameter of the pulley. That is, it allows the belt to ride closer to the hub or allows the belt to ride farther out from the hub. When more adjustment is needed than the adjustable pulley will give, another size of motor pulley must be used.

When adjusting a blower for cfm, you should make the necessary changes to the direct-drive motor connections or to the pulleys on a belt-drive unit. Then, recheck the cfm to get the air output as close to the desired output as possible. While the motor is running, also check the amperage draw to make sure that the motor is not overloaded. If the actual amperage draw on the motor exceeds the rated amperage rating of the motor, as given on the motor nameplate, the motor should be changed to a larger one.

18.3

Air Ducts

The **air ducts** in the distribution system carry the conditioned air from the unit to the rooms or areas in the building that are to be cooled or heated. The ducts are constructed of galvanized iron, aluminum, or fiberglass, and in all cases, they should be insulated. A typical duct system is shown in Figure 18–8. The air distribution

Air ducts

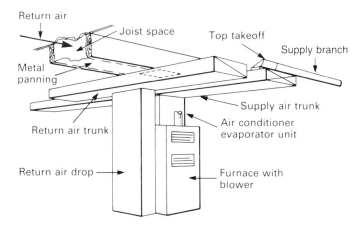

FIGURE 18–8
Blower and air ducts

ducts are sized to carry the cfm of air required to condition the building, which is also the air output of the cooling equipment. The system designer is responsible for the design of the duct system and for the sizing of the ducts. But in servicing a system, the technician may have to verify that sizing.

The layout or design of the air supply system may vary in different buildings. There are three main types of air distribution systems and many variations of each. The main types are perimeter loop, radial, and trunk-and-branch. Regardless of the layout used, the function of the system is the same, namely, to distribute the cooled air throughout the building. The proper distribution of air depends on the proper sizing of the ductwork.

Velocity method

One of the best ways for a service technician to check the duct sizing in a section of duct is by the **velocity method**. This method is based on the assumption that the velocity of the air in a properly sized duct will not exceed a set limit, measured in feet per minute (fpm). The allowable velocity for each type of duct is given in Table 18–1.

In the velocity method of duct checking, the duct is measured, the cross-sectional area is calculated, and

Applications	Residences	Schools, theaters, public buildings	Industrial buildings
Velocities for total face area			
Outdoor air intakes	500 (800)	500 (900)	500 (1200)
Filters	250 (300)	300 (350)	350 (350)
Heating coils	450 (500)	500 (600)	600 (700)
Velocities for net free area			
Air washers	500 (500)	500 (500)	500 (500)
Suction connections	700 (900)	800 (1000)	1000 (1400)
Fan outlets	1000–1600 (1700)	1300–2000 (1500–2200)	1600–2400 (1700–2800)
Main ducts	700–900 (800–1200)	1000–1300 (1100–1600)	1200–1800 (1300–2200)
Branch ducts	600 (700–1000)	600–900 (800–1300)	800–1000 (1000–1800)
Branch risers	500 (650–800)	600–700 (800–1200)	800 (1000–1600)

Note: Maximum duct velocities appear in parentheses.

TABLE 18–1   Recommended Duct Velocities for Various Applications

the velocity of the air is calculated from the area and the cfm. The cfm is determined from the size of the air conditioning or heat pump unit used. This method of calculating air velocity in a duct will not be entirely accurate because it does not take into consideration the resistance to the airflow in the duct. But it is accurate enough for fieldwork.

The steps in the velocity method are as follows. First, the cfm in the duct is found from the size of the equipment used. Next, the duct area is calculated. Finally, the velocity of the air is calculated by use of the following formula.

$$V = \frac{Q}{A}$$

where
V = velocity of the air, in feet per minute
Q = quantity of the air, in cubic feet per minute
A = area of the duct, in square feet

EXAMPLE

Suppose 2100 cubic feet per minute of air are flowing in a duct. The duct is 34 inches by 12 inches on the outside, with 1 inch of insulation inside. Find the velocity of air in the duct. Determine, from Table 18–1, whether that size is correct for a main trunk.

Solution

Since the duct is 34 inches by 12 inches on the outside, with 1 inch of insulation, it is 32 inches by 10 inches on the inside. The cross-sectional area, in square feet, is as follows:

$$A = \frac{32 \times 10}{144} = 2.22 \text{ ft}^2$$

Since the cfm is given as 2100, then the velocity can be found by using the formula.

$$V = \frac{2100}{2.22} = 945.95 \text{ fpm}$$

Reference to Table 18–1 shows that this velocity is within the limits for a main trunk. So, the duct sizing is correct.

If the quantity of air in the duct is correct for the part of the building served by the duct, and if the velocity is within the limits given on the velocity chart, then a technician can assume that the section of duct will work correctly. If the duct is too small, then the engineer or designer of the system should be notified.

Return air ducts can be checked in the same way as supply ducts. Remember that it is just as important for the return air ducts to be properly sized as it is for the supply ducts.

18.4

Supply Registers and Return Air Grilles

Supply registers

Supply registers are located in each space that is cooled or heated by the air conditioning or the heat pump system. The registers are connected to the air supply ducts, and they are the terminal devices in the supply system. They are sized to allow the proper amount of air to flow into the room, according to the system design. In addition, registers are shaped to discharge the air in a particular pattern, according to the shape of the room.

Figure 18–9 shows a typical wall register. Most supply registers, such as the one shown in Figure 18–9, have balancing dampers built into them. These dampers are used to adjust the quantity of air leaving the register.

Supply registers must be sized, shaped, and located so that they provide even air distribution in a building and provide the proper amount of air to each part of the building. A complaint of uneven cooling in a building may be the result of improper location or sizing of the registers. An air conditioning or heat pump serviceperson should check the registers when such a complaint is received.

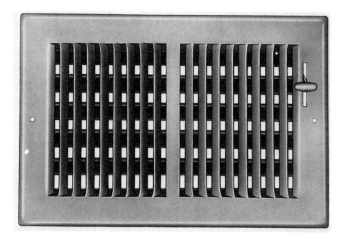

FIGURE 18–9
Supply register (Courtesy of Lima Register Co.)

Checking Supply Registers

To check a supply register for proper cfm requires the use of a **velometer** or **anemometer**. These instruments are used for measuring airflow rates. Using either instrument, read the velocity of the air coming from the register. Calculate the cfm by use of the following formula.

Velometer

Anemometer

$$Q = V \times A$$
where
Q = quantity of the air, in cubic feet per minute
V = velocity of the air, in feet per minute
A = free area of the register face, in square feet

EXAMPLE

The measured velocity of the air at the face of a register is 320 feet per minute. The free area of the register is 202 square inches. Determine the quantity of air flowing out of the register.

Solution

Find Q by using the formula.

$$Q = 320 \times \frac{202}{144} = 448.89 \text{ cfm}$$

The pattern of the air coming out of the register is just as important as the cfm. Each register distributes air in a particular pattern. Air leaves a register at a fairly high velocity, but the air's effectiveness for cooling or heating only extends to a point where the velocity is 50 fpm or greater. This terminal velocity defines a curtain of air, called an **envelope**, over which the register has effect. The distance the envelope extends directly out from the register is called the **throw** of the register. The width of the envelope is called the **spread**.

The envelope of air leaving a register is fan-shaped. It should cover the windows, the doors, and any other part of the room that adds to the heat gain or loss of the room. This pattern is shown in Figure 18–10. In checking the supply register, you should observe the type of register used. Then, refer to the register manufacturer's catalog for information about the spread and the throw of the air from the register in order to deter-

Envelope

Throw
Spread

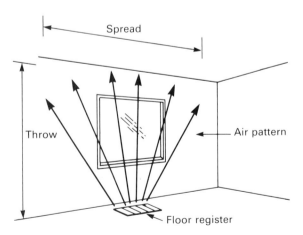

FIGURE 18–10
Air pattern from supply register

mine the pattern of the air. If the cfm or the air pattern is not correct for an application, the register should be changed to provide correct performance.

Checking Return Air Grilles

Just as supply registers must be sized, shaped, and located so that they provide even air distribution, return air grilles must be properly sized and located for distribution of the return air. Figure 18–11 shows a typical return air grille. The **return air grilles** in a system should be large enough to return the same amount of air as the supply registers provide.

    The procedures used for checking supply registers can also be used for checking the size of return air grilles.

Return air grilles

FIGURE 18–11
Typical return air grille (Courtesy of Lima Register Co.)

A physical check of the supply system, or reference to the mechanical plans of the building, will show how much air should be carried by the return air grilles. That amount will be the total cfm of the supply registers in the part of the building in which the return air grilles are located.

18.5

Dampers

Damper

A **damper** is a device used to control the volume of air flowing through or from an orifice. Most air dampers are made of sheet metal, and most are in the form of an adjustable gate or louver that can be opened or closed by a lever located on the outside of the damper. A **balancing damper** is a damper installed inside a section of supply duct, and the adjusting lever extends outside the duct. A typical balancing damper is shown in Figure 18–12. Balancing dampers are installed in each branch of a duct system. They are used to ensure that the proper amount of air flows into each branch.

Balancing damper

Some supply air registers have dampers built into them. When dampers are located in the ducts and also in the registers, the dampers in the registers should be set fully open. Sometimes, there is more than one register on a duct with a balancing damper in it. In such cases, the duct damper is used to provide the required air for the entire section of duct. The register dampers are used to balance the airflow at each register.

Another type of damper is used to isolate sections of a duct in order to shut off the flow of air in case of fire. These dampers are called **fire dampers**. Fire dampers have springs on the blades, as shown in Figure 18–13. The springs hold the blades closed. The dampers are held open against the spring's tension by a metal link attached to the damper blades. The link is made of fusible metal; that is, it will melt if it gets hot. If the air in the duct or around the duct gets very hot, as it will in a fire, the link melts and the spring pulls the damper closed. With the damper closed, the air distribution system will not spread the fire.

Fire damper

If a complaint is received about an insufficient supply of air to a space, the balancing dampers in the system

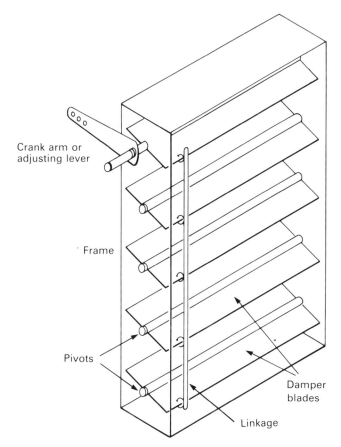

Crank arm or
adjusting lever

Frame

Pivots

Damper
blades

Linkage

FIGURE 18–12
Balancing damper

should be checked. Make sure that they are set for the proper amount of air for each branch or register. Make sure that any fire dampers in the system are open. The dampers can be checked visually or manually.

18.6

Summary

The air distribution system is the part of the air conditioning or the heat pump system that distributes conditioned air through the building. The proper functioning

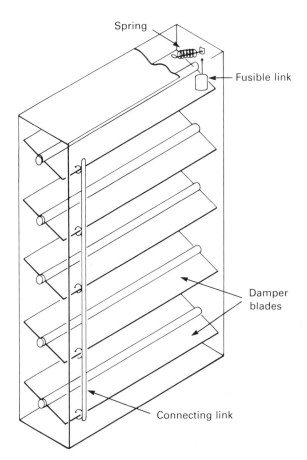

FIGURE 18–13
Fire damper

of the air distribution system is as important to the total operation of the system as is the functioning of any other part of the system.

The air distribution system consists of the supply blower, the supply and return air ducts, the supply registers, the return air grilles, and the dampers or any other air control device. If improper air distribution is affecting the comfort of the people in a building, then each of these parts of the system may have to be checked in order to determine the problem.

In a packaged air conditioning or heat pump system, the blower is located in the unit itself. In large systems, the blower may be located in one part of the duct system, and the evaporator or indoor coil may be located in another part. Air flows from the blower through the evaporator or indoor coil and then through the ductwork. The blower must be sized and adjusted properly so that it will supply the right amount of air for the proper operation of the equipment and the cooling of the building.

The supply and return air ducts, the supply registers, and the return air grilles must all be properly sized to carry the cfm required for the system. Improperly sized or located ducts or registers should be reported to the system engineer, who can then replace them with the proper size.

Some systems have balancing dampers in the branch ducts or in the main trunks. Building codes for large buildings require fire dampers in the ductwork between main sections of the building. In servicing an air distribution system, a technician should check all dampers in the system to be sure that they are set for the proper airflow.

18.7

Questions

1. Describe the term *forward-curved* in relation to a blower used in an air distribution system.
2. Name two things that can affect the cfm output of a blower.
3. Name three things that can cause a complete lack of airflow in a distribution system.
4. Describe two different ways to check the quantity of air flowing across a coil.
5. Is the following statement true or false? The pressure drop across all coils is the same.
6. On a belt-drive blower system, how do you change the cfm of the air being delivered?
7. How do you change the cfm of the air being delivered on a direct-drive blower?
8. A system is sized to deliver 36,000 British thermal units per hour. The temperature drop across the coil is 18.6 degrees Fahrenheit. Determine how much air is moving across the coil, measured in cubic feet per minute.

9. A centrifugal blower delivers 1260 cubic feet per minute of air when it is operating at 960 revolutions per minute. Determine the cfm it will deliver if it is set to turn at 850 revolutions per minute.

10. Is the following statement true or false? The velocity of the air flowing in a given size of duct is directly related to the quantity of the air flowing.

11. The air in a 12-inch by 10-inch duct is moving at 780 feet per minute. Determine the amount of air that is flowing, in cubic feet per minute.

12. Is the following statement true or false? If the duct system in an air distribution system is sized correctly, the sizes of the supply registers and the return air grilles are not important.

13. Is the following statement true or false? If an air distribution system has properly sized supply registers, then the return air grilles can be much smaller than these registers.

14. Name the devices used in duct branches to adjust the flow of air to the registers.

Heat Pump System

19.1

Introduction

A **heat pump** is a mechanical refrigeration device that can provide either air conditioning or heating in a building. Which of the two is actually provided depends on the direction in which the refrigerant flows through the system. The direction of the flow of the refrigerant is a function of the control system used with the unit.

A heat pump does not produce cooling or heating. It simply moves heat from one place to another. During a cooling cycle, a heat pump picks up heat from the air inside a building and discharges it outside the building. During a heating cycle, heat is picked up from the air outside the building and discharged into the building.

This chapter describes the heat pump system, including its mechanical design, the heat source, and the parts and controls specifically used on heat pumps.

Heat pump

19.2

Mechanical Design

A heat pump is basically an air conditioning system. It has additional devices and controls for reversing the direction of the refrigerant flow and, consequently, the function of the evaporator and the condenser. The reversal of the refrigerant flow is accomplished by the use of a special valve called a reversing valve. Figure 19–1

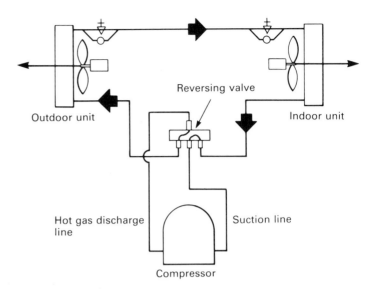

FIGURE 19–1
Schematic of heat pump operating in cooling mode

shows a schematic diagram of a heat pump system in the cooling mode.

To understand how a heat pump works, you must understand how a refrigeration system is used in an air conditioning unit. Chapters 5 and 6 in this text cover the general operation of refrigeration systems and air conditioning systems. A review of those two chapters will help you understand how a heat pump system works.

In an air conditioning system, the refrigerant flows from the outlet of the compressor to the condensing section of the system. In a heat pump, the refrigerant leaves the compressor and goes to one connection on the reversing valve. Depending on whether the valve is in the heating mode or the cooling mode, the refrigerant leaves the valve in one of two lines. One line runs to what is normally the evaporator coil, and the other line runs to what is normally the condenser coil. In the **cooling mode**, the refrigerant goes from the reversing valve to the condenser or outside coil, as shown in Figure 19–1. It returns through the evaporator or inside coil.

If the reversing valve is moved to the **heating mode**, then the direction of the flow of refrigerant to the two

Cooling mode

Heating mode

coils is reversed. The refrigerant now goes from the reversing valve to the indoor coil, as shown in Figure 19–2. It returns through the outdoor coil. The functions of the two coils are also reversed. The coil that was an evaporator becomes a condenser, and the coil that was a condenser becomes an evaporator. When the unit is operating in the reversed position, heat is picked up by the refrigerant in the outside coil and is rejected inside.

19.3

Heat Source

A heat pump, unlike a typical heating unit, does not produce heat. It simply transfers heat from one place to another. So, the use of a heat pump is practical only if heat is available for transfer. Any material with a temperature above 0 degrees absolute has heat in it. But the higher the temperature of the material, the better source of heat it is for the heat pump.

     Groundwater and outside air are the two most common sources of heat for typical heat pumps. **Ground-**

Groundwater

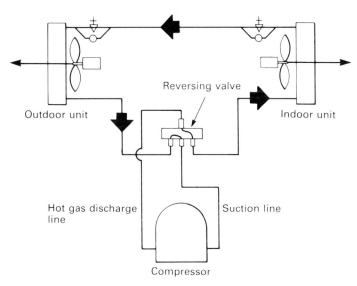

FIGURE 19–2
Schematic of heat pump operating in heating mode

Outside air

Heat sink

Auxiliary heat

water is usually found at a temperature of around 45 to 50 degrees Fahrenheit. It is almost always a good source of heat for a heat pump. **Outside air** is a good heat source any time that the air temperature is above about 15 degrees Fahrenheit. The heat source, either water or air, is called a **heat sink**. In an air source heat pump, the outside air is blown directly across the outside coil of the unit. In this text, only the air-to-air heat pump is covered in detail. But many of the service procedures described are the same for other types of heat pumps.

19.4

Auxiliary Heat

The heat output of an air-to-air heat pump varies as the outdoor temperature varies. When the outdoor temperature is warm, more heat is available for the heat pump to move inside. So, the heat output of the unit is greater. When the outdoor temperature is cool, less heat is available, and the heat output of the unit is less. Consequently, a heat pump produces the least amount of heat, because of the colder outside temperature, when the heat loss of the building is the greatest. As compensation for the lower heat output at lower outdoor temperatures, extra heating capacity is provided by some form of **auxiliary heat**. The auxiliary heat can be produced by any type of conventional heating system, but it is most commonly provided by electric heat.

When electric heat is used as the auxiliary heat source, it may be in the form of a conventional electric furnace installed with the indoor section of the heat pump. Or it may be in the form of electric heating elements that are part of the indoor blower unit of the heat pump. In either case, the auxiliary heat components are sized to provide any heat for the building that the heat pump cannot provide. The auxiliary heat units are on only when the temperature outside is too low to provide a good source of heat for the heat pump.

The heat pump control system is designed to turn on the auxiliary electric heat if the heat pump cannot satisfy the thermostat on a call for heat. So that energy is saved, an outdoor thermostat is often used in conjunction with the indoor thermostat. The outdoor thermostat keeps the electric heat turned off until the outdoor temperature gets down to the balance-point temperature for the building. The balance-point temperature is the temperature at which the heat pump's output just balances the heat loss of the building.

19.5

Parts and Controls

A heat pump system has some special parts and controls that are not found on air conditioning systems. The three parts common to all heat pumps are reversing valves, special cooling/heating thermostats, and defrost controls. These parts are described in the subsections that follow.

Reversing Valves

The main part of the operating system of a heat pump is the **reversing valve**, such as the one shown in Figure 19–3. This valve is used to reverse the flow of the refrigerant in the system to produce both a cooling and a heating operation. A reversing valve has connections for four refrigerant lines. One connection is for the hot gas discharge line from the compressor. One connection is for the suction line to the compressor. The other two connections are for the lines that go to one or the other of the system coils.

Reversing valve

A reversing valve is a cylinder with a free-moving piston inside. See Figure 19–4. The piston can move from one end of the cylinder to the other. When the piston is positioned at one end of the cylinder, refrigerant flowing into the valve through the discharge line from the compressor is directed out of one line to one of the coils. When the piston is positioned at the other end of the cylinder, refrigerant that enters the valve through the same inlet port is discharged to the opposite coil. In either case, the suction line to the compressor draws refrigerant from the line that is opposite the line used for discharge.

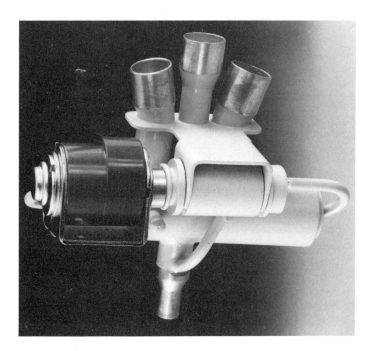

FIGURE 19–3
Reversing valve (Photo courtesy of ALCO Controls.)

Heat Pump Thermostats

Heat pump thermostat

A **heat pump thermostat** must incorporate the characteristics of a system switch, a heating thermostat, and a cooling thermostat. It must be able to switch the system from a cooling mode to a heating mode and to control the operation of the system in either mode. So that it can perform these operations, the thermostat is specially designed and wired into the control circuits. One of its circuits controls the operation of the reversing valve, another circuit controls the operation of the compressor in the cooling mode, and another circuit controls the compressor and a source of auxiliary heat in the heating mode.

Many different types of thermostats are used for controlling heat pumps. Some manufacturers use single-bulb thermostats with a manual switch for heating/cooling mode control. Others design their heat pump control systems so that a multibulb, multistage thermostat must be used.

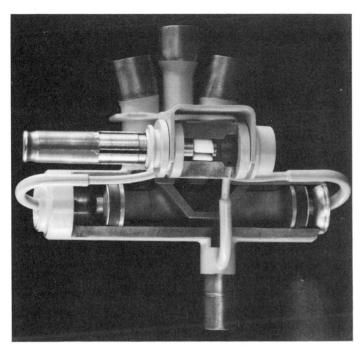

FIGURE 19–4
Cutaway view of reversing valve (Photo courtesy of ALCO
Controls.)

If a **single-bulb thermostat** is used, it must have a
dual-contact bulb. This type of bulb has a set of heating
contacts in one end and a set of cooling contacts in the
other. The reversing valve is controlled by a manual
system switch. Operation of the compressor for heating
and cooling is controlled by the contacts in the bulb.
These circuit controls are illustrated in Figure 19–5, which
shows a schematic diagram of a heat pump system with
a single-bulb thermostat.

When a **multibulb thermostat** is used, the system
can include automatic changeover from heating to cooling if desired. With a multibulb thermostat, normally
one or two stages of heating are controlled by one or
two mercury bulbs. Also, one or two stages of cooling
are controlled by other bulbs. The reversing valve is
controlled either by one of the heating bulbs or by one
of the cooling bulbs. Operation of the compressor for

Single-bulb thermostat

Multibulb thermostat

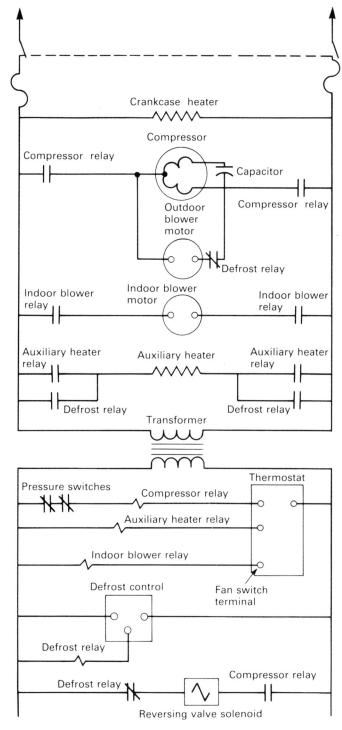

FIGURE 19–5
Wiring diagram of heat pump system using thermostat with single bulb

heating and cooling is then controlled by the other bulbs. These circuits are indicated in Figure 19–6, which shows a typical wiring layout for a four-bulb thermostat controlling a heat pump.

Defrost Controls

When a heat pump is operating in the reverse, or heating, cycle, the outdoor coil of a heat pump functions as an evaporator. In this mode, the coil temperature will often be below the freezing temperature of water. Moisture in the air freezes on the coil surface as outdoor air is drawn across the coil. So that this frozen moisture does not build up to the point where it blocks the flow of air across the coil, it must be melted. A special control, called a **defrost control**, is used for this melting. Figure 19–7 is a schematic diagram of a typical defrost control system.

Defrost control

A defrost control is used to initiate an action that will melt the frost from the outside coil and then terminate that action when the defrosting is complete. Several methods are employed for this task, but the most common method is **hot gas defrost**. In a hot gas defrost system, the heat pump, which is in the heating mode, is temporarily switched into the cooling mode. At the same time, the outdoor coil blower is turned off. Simultaneously, the auxiliary heat system is turned on to prevent cold air from being blown into the building during the defrost cycle. Switching the unit to the cooling cycle causes the outdoor coil to be flooded with hot discharge gas from the compressor, and the coil is quickly defrosted. The defrost control system initiates the defrost cycle when it is needed and terminates it when the frost or ice is melted off the coil.

Hot gas defrost

A defrost control system must have a sensing device to determine when the outside coil is beginning to frost up, a process called **initiation**. And it must be able to sense when the coil is free of ice, a process called **termination**. There are several ways in which a defrost system can be initiated and terminated. Among the methods commonly used are time, temperature, and refrigerant pressure.

Initiation

Termination

When the **time method** is used to initiate and terminate a defrost cycle, a clock is employed. The clock

Time method

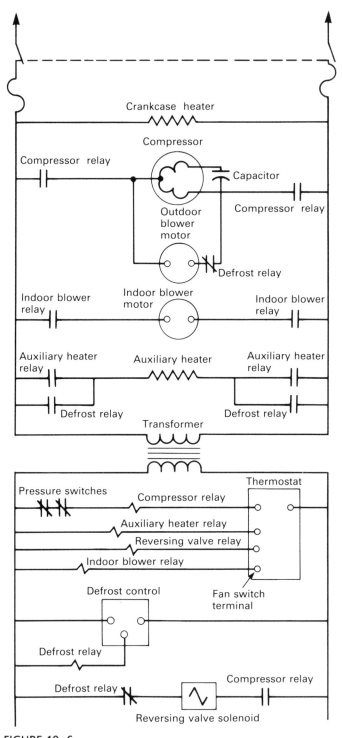

FIGURE 19–6
Wiring diagram of heat pump system using thermostat with four bulbs

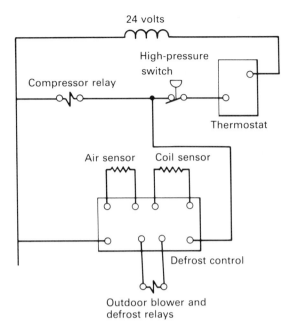

24 volts

High-pressure switch

Compressor relay

Thermostat

Air sensor   Coil sensor

Defrost control

Outdoor blower and
defrost relays

FIGURE 19–7
Defrost control system

calls for a defrost cycle at a predetermined time interval, and it terminates the defrost cycle after another predetermined time interval. Both the starting cycle time and the defrost period time are set according to the temperature and humidity conditions of the location of the system.

The **temperature method** of determining the proper time to start and stop a defrost cycle uses the air temperature entering and leaving the outdoor coil. If frost or ice starts to build up on the coil, the temperature difference across the coil will decrease. The decrease in temperature is sensed by a thermostatic control mounted on the inlet side of the coil and another control mounted on the discharge side. These controls feed information to a device that monitors the differential between them. If the differential is reduced, indicating a need for defrosting, a defrost cycle is initiated. As the frost or ice on the coil is removed by the hot gas, the temperature difference across the coil increases. This increase is sensed by the two thermostats. Hence, the thermostat control sends a signal to the defrost control to terminate the cycle.

Temperature method

Refrigerant pressure method

A third commonly used defrost control sensing system is the **refrigerant pressure method**. Since the pressure in the coil is representative of the temperature of the refrigerant, the coil temperature can be monitored closely by monitoring the refrigerant pressure. A pressure tap is placed in one of the refrigerant lines in the coils, usually on a return bend, and a tube from the tap leads to a pressure switch. If the coil starts to frost up, the pressure of the refrigerant in the coil will drop. If it drops low enough to indicate a freezing temperature, the pressure switch activates the defrost control.

While defrost controls are necessary for the proper operation of a heat pump system, they are also the source of some service problems. A trained heat pump troubleshooter or serviceperson must be able to troubleshoot and service the defrost controls as well as the other parts of the system.

Auxiliary Heat Controls

The control for the auxiliary heat on a heat pump system is basically a function of the thermostat. When the thermostat calls for heat, the reversing valve is already in the heating mode position. The first-stage heating bulb of the thermostat operates the compressor. When the reversing valve is in the heating position, the heat pump provides heat. If the heat output of the heat pump is not sufficient to bring the temperature in the building up to the thermostat's set point, a second-stage heating bulb calls for auxiliary heat. The auxiliary heating elements are energized when the second-stage bulb energizes the heating relay.

The heat pump should do as much of the heating as possible; the electric heat should only be used when it is absolutely necessary. Therefore, an outdoor thermostat is often used in conjunction with the regular thermostat in order to control the auxiliary heat. On the outdoor thermostat, a set of contacts closes when the temperature is below the set point. The contacts in the outdoor thermostat are wired in series in the control circuit that energizes the heat control relay. This method of wiring keeps the electric elements from being energized until the outdoor temperature is below the set point temperature.

If necessary, the auxiliary heat may also be controlled by the defrost control system. When the heat pump goes into a defrost cycle, the auxiliary heat should come on to heat the air that is blown into the building.

19.6

Summary

A heat pump is an air conditioning system that can be operated in the normal mode for cooling the air and in the reverse mode for heating the air. In the normal, or cooling, mode, a refrigerant is circulated through a closed-loop piping system. It picks up heat from the air inside a building and discharges the heat outside the building. In the reverse, or heating, mode, the same refrigerant picks up heat outside the building and discharges it inside the building.

The direction of the refrigerant flow in the system determines whether the unit operates as an air conditioning unit or as a heating unit. The direction of the refrigerant flow is controlled by a valve called a reversing valve. The reversing valve is controlled by a thermostat located in the building.

When a heat pump is operating in the heating mode, the temperature of the outside coil may often be lower than the freezing temperature of water. As moisture condenses on the coil from the cool air passing over it, frost and ice form on the coil. So that the frost and ice do not build up to a point where they will affect the operation of the unit, a defrost control is used. This control melts any accumulation of frost or ice from the coil.

19.7

Questions

1. Explain the following statement: "A heat pump does not produce heat; it merely moves it from one place to another."
2. Is the following statement true or false? A heat pump is an air conditioning system in which the direction of the flow of refrigerant can be reversed.

3. What function does a reversing valve perform in a heat pump?

4. Describe the terms *cooling mode* and *heating mode* for a heat pump system.

5. Why is it necessary to furnish a source of auxiliary heat with a heat pump system?

6. What is the most common source of energy for auxiliary heat?

7. How does an outdoor thermostat save energy when it is used with a heat pump system?

8. Name three common parts and controls used on heat pumps that are not used on air conditioning systems.

9. What happens to the refrigerant flow through the heat pump system when the piston in the reversing valve moves from one end to the other?

10. What is the function of a system switch on a heat pump thermostat?

11. Why is it necessary to have both cooling and heating bulbs in a heat pump thermostat?

12. What causes frost or ice to build up on the outdoor coil of a heat pump during a heating cycle?

13. Describe how a hot gas defrost cycle removes frost or ice from the outdoor coil.

14. What do the terms *initiate* and *terminate* mean when used in relation to a defrost system?

Service Procedures Common to All Heat Pump Systems

CHAPTER

20

20.1

Introduction

Many service procedures used for air conditioning systems are also common to heat pump systems. Most of these procedures are related to the refrigeration components. Those procedures that require special consideration when a heat pump is being checked are discussed in this chapter.

In a heat pump system, a refrigerant is circulated through a refrigeration system to carry heat from the low-pressure side to the high-pressure side. The pressures in the two sides of the system are reversed as the system is changed from the cooling mode to the heating mode of operation. Regardless of the mode of operation, problems in the high-pressure side of the system affect the low-pressure side. Similarly, problems in the low-pressure side affect the high-pressure side. Problems in the refrigeration system are usually indicated by pressure or temperature variations of the refrigerant from those expected for normal operation. Common procedures for checking for problems within the refrigeration system include checking pressures and temperatures.

There are some service procedures that are common to all heat pump systems. They are generally related to the refrigeration system—more specifically, to the refrigerant pressures in the system. These common procedures, which are presented in this chapter, include checking refrigerant pressures, checking the refrigerant charge, leak testing, pumping down a system, evacuating a system, and charging a system. Any one of these tasks may have to be performed regardless of what part of the system is being checked. Since they have to be performed often, a service technician should be familiar with each procedure.

20.2
Checking Pressures

There are a number of problems that may occur in a heat pump system that will cause high or low pressures in the system. The main causes of high pressure in a system are an overcharge of refrigerant, a reduction in the amount of air across the condenser coil, a dirty condenser coil, or noncondensable gases in the condenser coil. Low pressure may be caused by a low refrigerant charge in the system or a low condenser air temperature. There are other causes of both conditions, but the causes just listed are the most common.

Gauge manifold

The pressures in a refrigeration system are checked by using a **gauge manifold**. The gauge manifold is a device with valves, gauges, and hose connections. These components are arranged so that it is convenient to check pressure, evacuate a system, or add refrigerant to a system.

A gauge manifold has two gauges mounted on it: a high-pressure gauge and a compound gauge. Each gauge is connected through internal porting to a hose connection, as shown in Figure 20–1. The hose connections are used to run hoses from the manifold to the service ports on the heat pump. Each hose connection, in turn, is connected internally to a center hose connection port. The connection between each of the gauge ports and the center port is controlled by a valve on the manifold.

To use a gauge manifold to check the pressures in a refrigeration system, open the disconnect switch so that the unit cannot turn on. As shown in Figure 20–2, connect the hose from the high-pressure gauge port to a service port on the high-pressure side of the system. Connect the hose from the low-pressure gauge port to

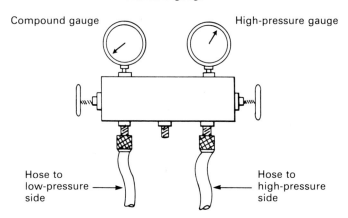

FIGURE 20–1
Manifold gauge connections

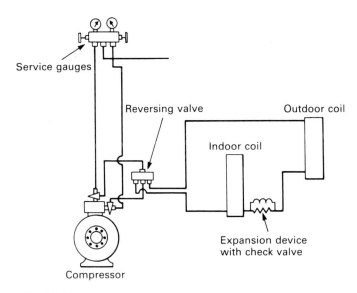

FIGURE 20–2
Service gauge connections for checking pressures in heat pump system

Sight glass

Liquid level test valves

Snifter valves

a service port on the low-pressure side of the system. If the service ports are located on service valves on the unit, backseat the service valves before making the hose connections. If the connections are made at Schrader fittings, then use a Schrader fitting coupling or a hose with a Schrader valve upsetting pin in it. Make sure that the valves on the manifold are closed so that refrigerant will not be lost from the center hose when the connections are made.

To open the system to the service gauges, turn the service valve stem in two turns. Use a valve wrench to turn the valve. When the gauges indicate pressure in each hose, purge the air out of the hose by slightly loosening the knurled nut connecting the hose to the manifold. When refrigerant flows from the loosened nut, close it again. Start the unit, and observe the pressures in the two sides of the system on the gauges.

20.3

Checking the Refrigerant Charge

Leaks in a system or improper charging may leave a heat pump system with too little or too much refrigerant in it. Since every system is designed to produce its rated cooling and heating output when it is correctly charged, a serviceperson must be able to properly check the charge and determine if it is the correct charge for the system.

There are several methods for checking the refrigerant charge in a heat pump system. The best method for any given unit depends on the unit's design. The three major methods for checking the charge are the use of sight glasses, the use of liquid level valves, or the use of pressure/temperature charts. These charts show the operating pressures for various condensing and evaporator air temperatures. Pressure/temperature charts for each unit are furnished by the equipment's manufacturer. Each of the three methods is described in the following subsections.

Sight Glasses

A **sight glass** is a fitting in the refrigerant line that has a glass window in it through which the refrigerant can be seen. A sight glass is illustrated in Figure 20–3. When it is used for checking refrigerant charge, a sight glass is usually installed in the liquid line where it leaves the condensing coil.

To check a unit that is equipped with a sight glass in the line leaving the indoor unit, make sure that the unit is operating in the cooling mode. Turn on the unit, and let it run for 5 minutes or longer so that the temperature of the air entering the condenser is about normal for the system. Observe the refrigerant in the sight glass.

With a proper refrigerant charge, or even with an overcharge, the sight glass will be filled with liquid refrigerant. If bubbles appear in the refrigerant flowing through the glass, there is vapor in the refrigerant. The bubbles indicate that all of the refrigerant is not condensing in the condensing coil. The bubbles are an indication of an undercharge or of some other problem in the condensing section. If the problem proves to be an undercharge, then more refrigerant should be added to the system.

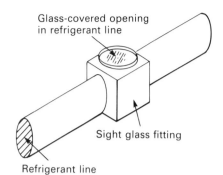

FIGURE 20–3
Sight glass

Liquid Level Test Valves

Liquid level test valves are small valves, sometimes called **snifter valves**, that are installed on refrigerant lines, coil return bends, or liquid receiver tanks. Figure 20–4 illustrates typical liquid level test valves. A snifter valve has a very small outlet opening. When the valve is opened, a sample of refrigerant flows out.

To use a liquid level test valve to check the liquid level in a unit, put the system in the cooling mode, and open the valve slightly. Observe the refrigerant leaving the valve. If the refrigerant is in the gaseous state, then only refrigerant vapor is present in the system at the location of the valve. If liquid leaves the valve, then the liquid level in the system is above the level of the valve.

If a unit has only one liquid level test valve, it will usually be located on a return bend of the condenser coil or on the side of the receiver tank. It will always be located at the point in the system where the refrigerant should be in the liquid state, but just below the point where vapor may be present, for a given set of operating

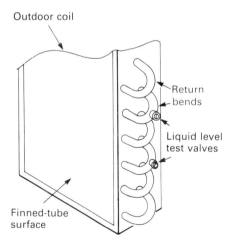

FIGURE 20–4
Liquid level test valves

conditions. To use the valve for checking the liquid level in the unit, operate the unit at the fixed set of conditions given by the manufacturer in the operating instructions manual. While the unit is operating, open the valve slightly at short intervals. If liquid only is emitted, the refrigerant level is above the valve. If vapor only is emitted, the refrigerant level is below the valve. If both liquid and vapor are emitted in spurts from the valve, the level of the refrigerant is just at the level of the valve.

If a system has two liquid test valves, they will be located at a high and a low position on the condenser coil return bends or on the side of the receiver. The lower valve is located on a part of the coil or tank where liquid should be present during normal operation. The upper valve is located where refrigerant vapor should be present during normal operation. To check the liquid level in the system, allow the unit to run long enough to reach stable operating conditions. Then, open each valve in turn. There should be liquid refrigerant only at the lower valve and vapor only at the upper valve. The liquid from the lower test port should be cold, and frost should form on the valve port while it is open. The vapor from the upper valve should be quite warm. If vapor is emitted from the lower valve, a low charge is indicated. If liquid is emitted from the upper port, an overcharge is indicated.

Pressure/Temperature Operating Charts

Pressure/temperature operating charts

Most manufacturers furnish **pressure/temperature operating charts** with their heat pump units. The chart shows the normal pressures that should be found in the high- and low-pressure sides of a system for various evaporator and condenser air temperatures. A typical pressure/temperature operating chart is shown in Figure 20–5.

To use a pressure/temperature chart for checking refrigerant charge, operate the unit at typical operating temperatures in the mode called for in the unit's manual. Check the pressures in each side of the system with pressure gauges. Compare the pressures displayed on the gauges with the pressures shown on the chart for a

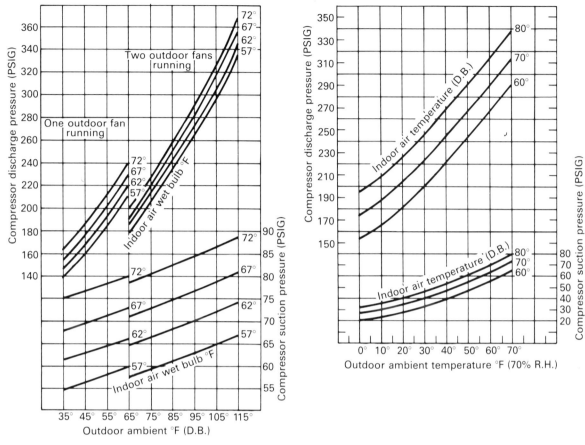

FIGURE 20–5
Pressure/temperature operating chart for heat pump

properly operating system at the same operating temperatures.

There is another way to check the refrigerant charge with a pressure/temperature operating chart. Attach an accurate thermometer to the lower section of the condenser grille, and measure the inlet air temperature. Slowly close the liquid line valve on the system until the low-side pressure is as shown on the chart for the refrigerant being used. Read the high-side pressure from the chart for that particular low-side pressure. If the system is properly charged, the actual pressures and temperatures should match those shown on the chart.

20.4

Checking the System for Leaks

If a leak is suspected in a refrigeration piping system, but you do not know what part of the system it is in, you must check the entire system for leaks. In this check, the entire system must be pressurized with refrigerant. However, if you know that the leak is located in a particular part of the system, you only need to pressurize that part of the system to search for the leak. If the unit is equipped with a liquid line valve and with compressor service valves, you can pressurize each side of the system separately.

If the entire system must be pressurized, you should do it through both the suction and the discharge service valves. In this case, pressure is applied to both sides of the expansion device. To apply pressure to the entire system, backseat both service valves. Attach the gauge manifold hoses to the service valve ports. Attach the refrigerant drum to the center port of the gauge manifold, as shown in Figure 20–6. Turn both service valves

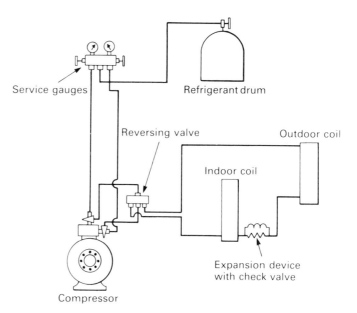

Service gauges

Refrigerant drum

Reversing valve

Outdoor coil

Indoor coil

Expansion device with check valve

Compressor

FIGURE 20–6
Service gauge connections for leak testing

to an intermediate position between open and closed. Open the drum valve, and open both valves on the gauge manifold. Now, refrigerant will flow into the system. Do not try to pressurize the system beyond a pressure corresponding to the saturation pressure of the refrigerant drum.

Check for leaks with an electronic leak detector, if one is available. If an electronic detector is not available, then a halide detector or soap bubbles can be used. Check all refrigerant lines and refrigerant line connections in the system for leaks. Pass the detector probe, or place the soap solution, along the lines and around each connector. Most refrigerants are heavier than air, so it is important to check on the bottom of the joints as well as on the top.

When all leaks are found, relieve the pressure in the system by releasing the refrigerant to the atmosphere. Or pump the refrigerant into a refrigerant drum used for waste refrigerant. Repair the leaks, and recharge the system with clean, dry refrigerant.

20.5

Pumping Down a System

A heat pump system with a liquid line valve and service valves on the compressor can be pumped down for service. **Pumping down** means to pump the refrigerant charge into the condensing section of the system through the use of the compressor. The liquid line valve and compressor discharge service valve can then be used to isolate the compressor, or the low-pressure side of the system. With the system pumped down, the compressor or the components and the parts in the low-pressure side of the system can be worked on without loss of refrigerant charge from the system.

To pump down a system, wire a toggle switch into the control circuit so that unit operation can be controlled. Figure 20–7 shows the location of the toggle switch when wired into the control circuit. Make sure that the thermostat is calling for operation in the cooling mode. Turn the unit off with the toggle switch. Lower the setting on the low-pressure switch to zero. Turn the

Pumping down

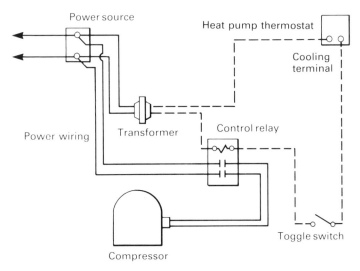

FIGURE 20-7
Toggle switch wired into control system for pumping down

thermostat up high enough so that it will continuously call for cooling. Set the fan switch on the thermostat to the on position.

Now, attach a gauge manifold to the gauge ports on the two service valves on the compressor. Attach the high-pressure gauge to the discharge valve service port, as shown in Figure 20–8. Attach the low-pressure gauge to the suction service port. Close the suction valve. Open the service valve gauge ports by turning the valve stems in one turn. Purge the air from the hoses.

Next, close the liquid line by frontseating it. This valve is located on the outlet of the receiver if the system has one, or it is in the liquid line between the condenser and the filter drier. Start the compressor by closing the toggle switch, and allow it to run until the suction pressure is reduced to nearly 0 pounds per square inch. Open the toggle switch to stop the compressor, and observe the suction pressure. The pressure will normally go back up somewhat at first. It should then level off.

Continue to cycle the compressor until the suction pressure remains at about 1 to 2 pounds per square inch when the compressor is not running. You may have to jumper the pressure switch with an electric jumper wire

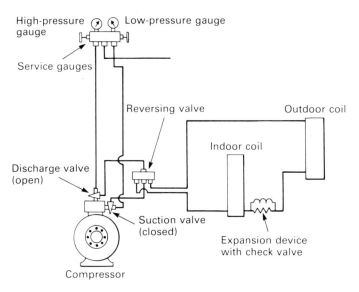

FIGURE 20–8
Service gauge connections for pumping down

to get the compressor to come on when the pressure gets down close to 0 pounds per square inch. Always keep a slight positive pressure in the system to prevent air and moisture from entering, in case there is a leak in the system.

When the suction pressure remains at about 1 to 2 pounds per square inch, frontseat the discharge service valve on the compressor to isolate the refrigerant in the condensing section of the system. The parts of the system from which the refrigerant has been removed—from the liquid line valve to the discharge service valve—can now be worked on. If any part of the system is opened to make any repairs, that part should be evacuated. A filter drier should be installed in the suction line before the refrigerant is allowed to enter that part of the system again.

To place the unit back in service, open the service valves. Slowly open the liquid line valve to allow the refrigerant to reenter the system. Reset the low-pressure switch to its normal operating pressure. Finally, remove the toggle switch, and check the system for proper operation.

20.6

Evacuating a System

If the refrigerant in a heat pump system becomes contaminated, the refrigerant in the system should be purged to the atmosphere. The system should be evacuated and then recharged with clean, dry refrigerant. The evacuation process is intended to remove any water vapor or air that may be in the old refrigerant.

If water is to be removed from the system, the pressure in the system must be low enough so that the water will vaporize. In this situation, a vacuum of 29.25 inches of mercury must be maintained when the ambient temperature is 70 degrees Fahrenheit. While the vacuum is being drawn, the pressure in the system should be measured with an accurate vacuum gauge. Either an electronic gauge or a mercury manometer should be used. The evacuation pump used must be capable of pulling a vacuum that low.

To evacuate the system, close all service valves. Attach a gauge manifold to the system so that the high-pressure gauge hose is on the liquid line valve. Attach the low-pressure gauge hose to the suction line valve. Use a Schrader valve if there is one on the suction line. Otherwise, use the valve port on the suction service valve. Install a vacuum gauge on the vacuum pump connection manifold. Connect the vacuum pump to the center port of the gauge manifold, as shown in Figure 20–9. Open the valves in the system and on the manifold, and start the pump.

Evacuate the system to at least 29 inches of mercury. Observe the vacuum gauge on the pump during the operation. If the process takes an abnormally long time, there may be a leak in the system. Stop the pump at least once during the initial evacuation cycle to determine whether there is a rapid loss of vacuum because of a leak. If the system will not hold a vacuum with the pump off, a leak is present. The leak must be repaired before you proceed.

When the evacuation process has been completed, close the manifold gauge valves. Stop the pump, and disconnect it from the manifold. If the unit is to be charged, or if refrigerant is going to be used to break

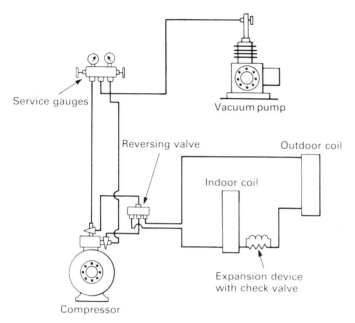

FIGURE 20–9
Service gauge and pump connections for evacuating a system

the vacuum in the system, the center hose can now be attached to a refrigerant drum. Several cycles of evacuation and purging with clean refrigerant will normally be required to thoroughly clean and dry a contaminated system. To repeat the evacuation cycle, break the vacuum with clean refrigerant by allowing enough refrigerant to enter the system to raise the pressure a few pounds. Then, repeat the evacuation process. Go through the evacuation process as outlined until you feel certain that the system is clean and dry. After the final cycle, the unit can be recharged.

20.7

Charging a System with Refrigerant

To add refrigerant to a heat pump system, attach a manifold gauge set to the system. The hose from the high-pressure gauge should be attached to a service port on the high-pressure side of the system. The hose from the compound pressure gauge should be attached to a service port on the low-pressure side of the system. The center hose from the gauge set should be attached to the drum of refrigerant used for charging. The connections here are the same as the connections shown in Figure 20–6.

Now, open the valve on the drum of refrigerant, and purge the hoses of any air they may have in them. Close the valve on the refrigerant drum after the hoses are purged, and start the unit. Open the valve on the high-pressure gauge of the manifold. The high side of the unit is now opened to the center hose. The valve on the low-pressure gauge remains closed. Open the valve on the refrigerant drum to allow the refrigerant to flow into the unit.

Note: If the pressure in the system is greater than the pressure in the refrigerant drum, the drum may have to be placed in a pan of hot water to increase the pressure so that the refrigerant will flow into the unit.

Watch the pressure gauges while the refrigerant is added to the unit. The refrigerant should be added in small quantities to prevent overcharging. Close the valve, and run the unit for several minutes after each addition of refrigerant to stabilize the operation of the unit.

Check the operating pressures against the pressures given for proper operation on the pressure/temperature chart to see whether the charge is correct. When the charge is correct, close the valve on the refrigerant drum. Backseat the service valves on the unit, and remove the manifold gauge set.

20.8

Summary

Some service procedures are common to all parts of a heat pump system. That is, they may have to be performed regardless of what part of the system is suspected of having problems. Most of these procedures relate to the refrigeration system. A service technician must be able to perform these service procedures properly, since the refrigerant charge is so important to the function of a unit. These common procedures include checking refrigerant pressure, checking the refrigerant charge, checking a unit for leaks, pumping down a system, evacuating a system, and charging a system with refrigerant.

The pressures on each side of a heat pump system in both modes of operation are directly related to the operation of the system. Checking these pressures during operation is an important method of determining how the unit is running. A manifold gauge set is used to check the pressures.

A proper refrigerant charge is necessary for proper operation of any heat pump system. Improper service procedures or leaks in the system can cause a system to be overcharged or undercharged with refrigerant. A service technician must know how to check the refrigerant charge in a system. Leaks in the refrigerant lines or in the connections on a system will allow refrigerant to leak out or allow air and water vapor to leak in. Being able to properly detect and repair leaks is one of the important tasks of a service technician.

If a unit has a liquid line service valve and service valves on the compressor, the refrigerant charge can be pumped down into the outdoor unit. In this way, a service technician can save the refrigerant charge while making repairs to the other parts of the system. This procedure often saves the cost of a new charge of refrigerant.

Evacuation of a system is necessary if the refrigerant gets air or water vapor in it. Evacuation is the process of drawing a low vacuum on the system so that water in the refrigerant will vaporize and can be removed with the refrigerant gas. A good-quality vacuum pump and vacuum gauge are necessary for this process.

If any leaks occur in the refrigerant line connections in a refrigeration system, some refrigerant is lost. Since the system must have the proper amount of refrigerant in order to operate the way it should, a service technician must know how to add refrigerant to a system. Any refrigerant added to a system must be dry, and no air can be allowed to enter with the refrigerant. Thus, a service technician must know how to charge a system properly when doing service work.

20.9

Questions

1. What are the main causes of high pressure in the refrigerant system of a heat pump?
2. What are the main causes of low pressure in a heat pump system?
3. Describe a gauge manifold.
4. How is the center port on the gauge manifold controlled in relation to the high-pressure gauge port or the manifold gauge port?
5. Name three common methods for checking the refrigerant charge in a system.
6. Describe a sight glass.
7. When a sight glass is used, what indicates a low refrigerant charge in a system?
8. How is a snifter valve used as a liquid level indicator in a refrigeration system?
9. Describe a pressure/temperature operating chart.
10. Name three methods for checking for leaks in a refrigeration system.
11. Is the following statement true or false? You do not need to remove the refrigerant from a system before applying heat to the fittings.
12. What is meant by *pumping down* a system?
13. What is meant by *evacuating* a system?
14. Describe how you charge a system.

Heat Pump Service Charts

21.1

Introduction

A **heat pump service chart** is a table or diagram showing the systematic steps to be followed when a technician is identifying a problem within a heat pump system. The process starts with a complaint concerning system operation. It leads progressively through the steps required to identify the individual part (or parts) that is causing the trouble. Proper use of a service chart will help a serviceperson locate problems in a heat pump system. This chapter describes the chart and explains its use.

Heat pump service chart

21.2

Description of the Chart

Service charts come in different forms. There are written discussions, graphs, and tables that describe the procedures a technician should follow when trying to find a problem in a heat pump. Each different type of chart is designed to help a service technician locate a problem in a system in the quickest and most efficient way. Most charts start with an obvious problem within the system. By observing how the equipment is functioning, and by analyzing what is observed, a service technician can eventually identify the part (or parts) that is faulty within the system through the use of the service chart.

A complete set of heat pump service charts is given in the Appendix at the back of this book. The charts are in tabular form. And they are referenced by the chapter in the text where service checks are described for the parts identified as possible problems. A portion of one of the charts is given in Table 21–1.

The heat pump service charts in this text are divided into three main sections. These sections are for complaints concerning the cooling-only operation, complaints concerning the heating-only operation, and complaints concerning both the cooling and the heating operation. In each chart, the first (left) column gives the general type of complaint. The complaints are no cooling or heating, no cooling, not enough cooling, too cool, no heat, not enough heat, too hot, noisy system, and too expensive to operate. Each chart also has five more columns. These five columns list the problem, the major section, the component, the part, and the reference. The following list describes the six column headings found in the service charts.

Complaint
- The **complaint** is a general statement about the operation of the equipment.

Problem
- The **problem** states how the system is performing.

Complaint	Problem	Major section	Component	Part	Reference
No cooling or heating	No air from supply registers	Electric power	Power supply	Branch: disconnect	Chap. 9
				Wiring	Chap. 9
				Fusing or breakers	Chap. 9
				Unit: disconnect	Chap. 9
				Wiring	Chap. 9
				Fusing or breakers	Chap. 9
			Control circuit	Transformer	Chap. 10
				Wiring	Chap. 10
				Thermostat	Chap. 22
				Control relay	Chap. 9
				Motor start relay	Chap. 12
				Motor overloads	Chap. 10
				Capacitors	Chap. 12
				High- or low-pressure switch	Chap. 10

TABLE 21–1   Portion of Heat Pump Service Chart for Cooling or Heating

- The **major section** describes the general part of the system that is affected.
- The **component** is the major part of the system affected within the section.
- The **part** is the individual part that is malfunctioning.
- The **reference** states the chapter in this text where the method of checking the part is described.

Major section

Component

Part

Reference

The first column on the left of the chart is a list of the general complaints that might be received about a system. The second column is a list of problems related to the specific complaint. The problems are listed in the order of their probability of being the cause of trouble in the system. The third column lists the major section of the system that is most probably at fault. The major sections are listed across from the problem that they are most closely related to, and they are listed in the order in which they should be checked.

The fourth column of the chart lists the components most likely to be at fault, and they are listed across from the major section in which they are located. Again, they are listed in the order of their probability of being at fault. The fifth column lists the parts related to the components, in the order in which they should be checked. The sixth (right) column shows the chapter where the service procedure for checking the parts is described. The entries in each column are shown across from the entries that they are related to in the other columns.

21.3

Using a Service Chart

To use the heat pump service charts presented in the Appendix, determine which chart is appropriate according to the complaint—that is, cooling only, heating only, or both cooling and heating. Next, locate in the chart the complaint received about the system. In the second column, identify the most probable problem from the information received in the complaint. In the third column, find the most likely major section of the system involved by analyzing the problem.

In the fourth column, find the most obvious component according to your investigation of the system. In the fifth column, identify the part that is probably at fault. Identification of the part may require a systematic checking of all the parts listed in the column across from the probable component. In the sixth column, locate the chapter in which a description is given for checking each part. The part can be checked by following the service procedures described in the chapter indicated.

An example showing the use of a heat pump service chart follows.

EXAMPLE

Assume that a complaint of "no cooling" during a cooling cycle has been received. Suppose also that initial investigations reveal that warm air is coming from the supply registers. Use the service charts in the Appendix to determine which component and part might be at fault.

Solution

Both the "cooling only" and the "cooling or heating" charts should be examined. The problem could be occurring during any cycle. We locate the charts for both of these conditions. And we find the complaint of "no cooling."

Figure 21–1 is a portion of the service chart showing the steps to be taken for this example, identified by number. The "no cooling" heading is listed as 1 on this figure.

In column 2, we find the problem "warm air coming from supply registers." This step is identified as 2 on Figure 21–1.

In column 3, four major sections are listed as possible sources of trouble. If the equipment seems to be running properly, but it just does not produce enough cooling, then the control system must be all right. If air is coming from the registers, the indoor blower is working. And since this component is in the low-pressure side of the system, we will assume that the low side is all right. Airflow also indicates that the air distribution system is operative. So, we only have the outdoor section of

Complaint	Problem	Major section	Components	Parts	Reference
No cooling Step one	Warm air coming from supply registers Step two	General	Refrigerant	Overcharge	Chap. 21
				Undercharge	Chap. 21
		Controls	Control circuit	Transformer	Chap. 10
				Wiring	Chap. 10
				Thermostat	Chap. 22
				Control relay	Chap. 9
		Outdoor section of the system Step three	Compressor	Motor	Chap. 12
				Valves	Chap. 15
				Start switch	Chap. 10
				Capacitor	Chap. 12
			Outdoor unit Step four	Blower motor Step five	Chap. 12 Step six
				Blower relay	Chap. 10
				High-pressure switch	Chap. 10
		Indoor section of the system	Indoor unit	Dirty coil	Chap. 17
				Low pressure switch	Chap. 10

FIGURE 21–1

Heat pump service chart with service steps numbered to match example

the system to investigate now. This step is shown as 3 on Figure 21–1.

In column 4, there are two components listed in relation to the high-pressure side of the system: the compressor and the outdoor unit. Visual observation should show us whether each component is running when it should be. Assume that this investigation shows that the outdoor blower motor is not running. Thus, the outdoor unit is the component at fault, as shown in step 4 in Figure 21–1. Also, this step identifies the part, as listed in column 5 of the chart and shown as step 5 in Figure 21–1.

Reference to the chapter listed in column 6 will give us a description of the procedures for checking the motor controls and the motor for proper operation. This step is shown as 6 in Figure 21–1.

21.4

Summary

Proper use of a service chart enables a service technician to identify problems within a heat pump system rapidly and accurately. Most charts are organized so that they lead a serviceperson through the checking procedure in a systematic fashion. They start from a broad and inclusive complaint about how the system is performing, and they lead to the identification of the individual part that is at fault. By carefully following the procedure outlined on a good chart, and by using the services procedures described in this text, a serviceperson can find problems within a system quickly and proficiently.

21.5

Questions

1. Describe a service chart.
2. Name the three main sections the heat pump service charts in this text are divided into.
3. Describe a complaint as related to servicing a heat pump.
4. Describe the general procedure for using any service chart.
5. How does the use of a service chart save time and money?

Checking Specific Parts of Heat Pump Systems

22.1

Introduction

A heat pump, as defined previously, is an air conditioning system that can produce both cooling and heating. Both modes are available in one system by reversing the direction of the flow of the refrigerant in the system. Many of the components and parts of a heat pump system are identical to those in an air conditioning system. But there are some additional components and parts that are related to the operation of the unit in the heating mode only. An air conditioning or heat pump service technician must be able to troubleshoot and service these parts as well as the common air conditioning parts.

The main parts of a heat pump that are not found on an air conditioning system are the reversing valve, the heat pump thermostat, the defrost controls, and special expansion device arrangements. These parts are directly related to the operation of the system as a heat pump. These parts and their function in the system were completely described in Chapter 19. This chapter presents the service procedures for checking each device.

22.2

Reversing Valves

The **reversing valve** in a heat pump is used to reverse the flow of refrigerant in the system for a heating cycle. The valve is basically a pilot-operated valve, with a pis-

Reversing valve

Free-floating piston

Pilot valve

ton that moves back and forth in a closed cylinder. A typical reversing valve is shown in Figure 22–1. A description of the parts of the valve and the procedures for checking the valve are given in the following subsections.

Connections and Parts of the Valve

The reversing valve has four connections for refrigerant lines, as shown in Figure 22–2. There are three connections on one side and one connection on the other side. On the side with three connections, the center connection runs to the suction line on the compressor. Each of the other two lines on that side goes to one of the coils. The line on the side that has only one connection goes to the discharge side of the compressor.

A **free-floating piston** in the main valve is positioned at one end or the other of the main cylinder by refrigerant pressure. The refrigerant pressure is controlled by a small **pilot valve** located adjacent to the main valve cylinder. The pilot valve is connected to the main valve cylinder by two very small capillary tubes, as shown in Figure 22–2. It also has a third small tube that connects it to the suction line from the compressor.

FIGURE 22–1
Heat pump reversing valve (Photo courtesy of ALCO Controls.)

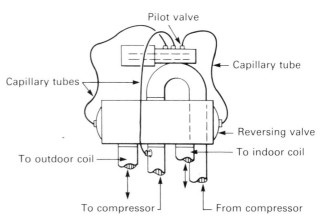

FIGURE 22–2
Refrigerant line connections to reversing valve

The pilot valve has a piston that is positioned by an electric solenoid. When the solenoid is not energized, the pilot valve piston is located at one end of its cylinder. Hence, the line from the compressor suction to one end of the main cylinder is opened, as shown in Figure 22–3A. If the solenoid is energized, the piston in the pilot valve moves to the other end. Then, the suction line going to the other end of the main cylinder is opened, as shown in Figure 22–3B. The main cylinder line that

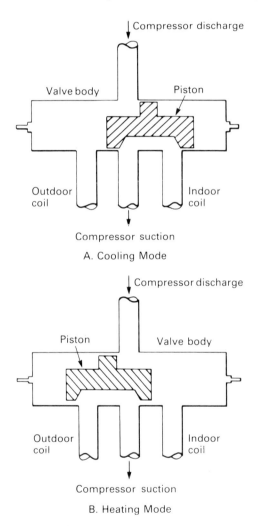

A. Cooling Mode

B. Heating Mode

FIGURE 22–3
Pilot valve piston positions for cooling and heating modes

is connected to the suction line in the pilot valve has a pressure that is lower than the main valve pressure. This pressure differential causes the main valve piston to move to the low-pressure end. Therefore, it switches the flow of refrigerant through the main valve.

There are three parts to check when a reversing valve is suspected of being at fault in a system. The three parts to check are the electric controls for the valve, the solenoid on the pilot valve, and the mechanical action of the valve itself. The electric controls should always be checked first. Service procedures for these three parts are described in the following subsections.

Checking the Electric Controls

A heat pump thermostat energizes the reversing valve when the operation changes from cooling to heating. To check the thermostat circuit to the reversing valve, look at the wiring diagrams for the main unit control. Determine whether the reversing valve solenoid should be energized in the cooling or in the heating mode. Then, set the thermostat so that the solenoid circuit should be energized.

Use a voltmeter, and set it for the control voltage used in the unit. Place the meter probes on the electric terminals where the control circuit connects to the solenoid coil, as illustrated in Figure 22–4. If there is no voltage at that point, then there is a problem in the thermostat or in the control power circuit to the thermostat. If the voltmeter shows the proper voltage for the control circuit, then the problem is in the reversing valve. The pilot valve solenoid should be checked next.

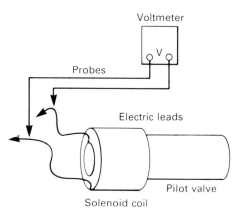

FIGURE 22–4
Checking electric power to solenoid coil of reversing valve

Checking the Pilot Valve Solenoid Coil

To check the pilot valve solenoid coil, make sure that the control circuit to the solenoid is energized. Do this check by following the procedure given in the previous subsection for the control circuit. The proper voltage for the control circuit should be indicated on the meter.

When the circuit is first energized, you should hear the solenoid piston move—that is, you should hear a clicking sound. You should also hear the refrigerant change direction of flow in the main valve. If you do not hear either sound, check the pilot valve solenoid coil by loosening the screw that holds it to the pilot valve. With the system energized, attempt to pull the coil away from the valve. *Caution:* Do not remove the coil entirely. As you move the coil, you should be able to feel the magnetic attraction of the coil for the solenoid stem if the coil is all right. If this test indicates that the coil is working, then remove one of the control wires from the solenoid. You should be able to hear the main valve reverse.

If the valve still does not seem to be working, then check the solenoid coil for continuity with an ohmmeter. While you have the control lead disconnected, place the ohmmeter probes on each of the coil terminals, as shown in Figure 22–5. If the coil has an open in it, then the ohmmeter will show infinite resistance. The coil is bad and will have to be replaced. If the ohmmeter shows that the coil is good, by indicating some reasonable resistance, then the problem must be within the pilot valve or the main valve.

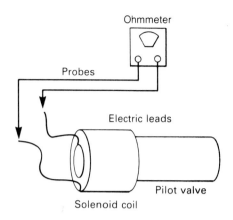

FIGURE 22–5
Checking solenoid coil with ohmmeter

Checking the Operation of the Reversing Valve

You can check the physical operation of the reversing valve by checking the temperatures of the refrigerant lines that run to and from the main valve. The refrigerant line running from the compressor discharge to the reversing valve should be hot to the touch. The line running from the valve to the suction side of the compressor should be relatively cool.

When the thermostat is calling for cooling, the refrigerant line running from the valve to the outdoor coil is at the same temperature (hot) as the hot gas discharge line from the compressor. And as shown in Figure 22–6A, the line running to the indoor coil is at the same temperature (cool) as the line from the valve to the suction side of the compressor. During a heating cycle, the direction of the refrigerant flow is reversed from the valve to the coils. So, the temperatures of the two lines to the coils are also reversed. That is, the line to the

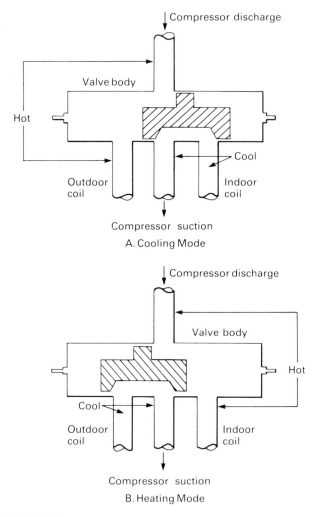

FIGURE 22–6
Relative temperatures of refrigerant lines for cooling and heating modes

indoor coil is hot, as shown in Figure 22–6B. And the line to the outdoor coil is relatively cool.

To check the operation of the valve, test the temperature of the lines in each mode. *Caution:* The hot gas discharge line can be hot enough to burn the skin. So, use a thermometer, or wear some protection for your hand when touching the line. If the temperature of the lines from the valve to the two coils does not change when the thermostat calls for a change of mode, something is physically wrong with the valve. It should be replaced.

If you must replace the reversing valve, use the same procedures that you would use for replacing any refrigerant line component. Remove the refrigerant from the unit. Make the change, being careful when soldering to protect the valve from getting hot. Evacuate the system before recharging.

22.3

Heat Pump Thermostats

A **heat pump thermostat** must energize the compressor on both the cooling and the heating cycle, energize the reversing valve to change between the modes of operation, and control the auxiliary heat used with the unit. There are many different types of thermostats used with heat pumps, and there are many different wiring arrangements for controlling the units. When servicing any particular system, carefully study the wiring diagram for that system to determine what control scheme is used for the unit. Then, check the thermostat and the electric circuits by using the procedures described in Chapter 10 for checking any thermostat.

Heat pump thermostat

22.4

Auxiliary Heat Controls

The auxiliary heat system provides any heat that the building needs but that the heat pump cannot produce because of very cold outside temperatures. The **auxiliary heat controls** on a typical heat pump system consist of a heating relay and either a combination of time delay relays or a sequencer. The time delay relays or the sequencer is used to bring the heating elements on the electric line in sequence rather than all at once. To troubleshoot or service the relays in the system, follow the procedures outlined in Chapter 10 for time delay relays.

Auxiliary heat controls

Sequencer

If a sequencer is used instead of relays, the various contacts on the sequencer should be checked as though each set were the contacts on a relay. The **sequencer** is actually a series of time delay relays set up to operate in the same way that individual relays wired together operate.

The electric power system to the elements is part of the regular power system for the heat pump. Each element is energized by a circuit that goes through a set of contacts in the relays or sequencer.

To check the power circuits, make sure that the thermostat is calling for auxiliary heat. Then, use a voltmeter to check the voltage at the elements. If there is no electric power at the elements when there should be, check across the relay or sequencer contacts with the voltmeter. If you get a voltage reading across the contacts, the contacts are open. If you get no voltage reading, the contacts are closed.

If there is an outdoor thermostat on the system, it must be set at a higher temperature than the existing outdoor temperature before the auxiliary heat will come on. To check the outdoor thermostat or the outdoor thermostat circuit, use the same procedures outlined for an indoor thermostat.

22.5
Defrost Controls

Defrost controls

Defrost controls are used on a heat pump to remove any frost or ice that collects on the outdoor coil because of the freezing of condensate during the heating cycles. There are several different types of defrost controls used by different manufacturers of heat pumps. In Chapter 19, the general types of defrost controls were described, and some methods of initiation and termination of defrost were covered. In this section, the electric components of a defrost control system are described, and methods of checking them for problems for initiation, and for termination are explained.

Components

On a call for defrost, a typical defrost control will do three things. It will position the reversing valve in the cooling position. It will turn off the outdoor blower. And

it will turn on the auxiliary heat unit. This control is
provided by one or more relays, called the **defrost re-**   Defrost relay
lay(s).

Either one electric three-pole relay or three individual one-pole relays are used. If a three-pole relay is
used, it has two normally closed (NC) sets of contacts
and one normally open (NO) set. If three relays are
used, two have NC contacts and one has NO contacts.
The NC contacts are wired to the control circuits to the
reversing valve and the outdoor blower motor. The NO
set of contacts is wired to a circuit that controls the auxiliary heat. Figure 22–7 is a wiring diagram for a typical
defrost control system.

Checking the Controls

If the outdoor coil begins to ice up but the unit does not
go into a defrost cycle, then the following checks should
be made. First, check the defrost relay coil. Make sure
that it is energized by checking across the terminals with

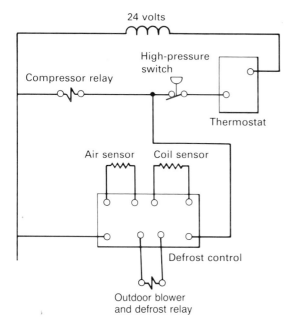

FIGURE 22–7
Wiring diagram for typical defrost control system

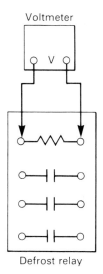

FIGURE 22–8
Checking defrost relay with voltmeter

a voltmeter. Figure 22–8 illustrates the connection of the voltmeter to the coil. A reading showing the correct control voltage for the system indicates that there is power to the coil. A no-voltage reading indicates that there is no power to the coil.

If you get a voltage reading, but the unit still does not go into defrost, disconnect one of the control circuit wires to the coil terminal. Check for continuity through the coil with an ohmmeter. Continuity is indicated by a reasonable resistance reading. Continuity shows that the coil is good. No continuity, or an open in the coil, is shown by a reading of infinite resistance. If the coil has an open, the coil is bad. The coil must be replaced if it is changeable. If the coil is not replaceable, the relay must be changed.

If the previous checks indicate that the relay coil is good, but there is still a problem in the defrost system, check the three sets of contacts in the relay (or relays). Check the voltage across the contacts with a voltmeter when the relay coils are energized. Open contacts show normal control voltage across the contacts. Closed contacts show no voltage. De-energize the relays and check again. You should be able to determine whether the contacts are opening and closing as they should. If they are not, the relays should be replaced.

Defrost Initiation

As explained in Chapter 19, there are several ways to initiate a defrost cycle. Three common methods are time, temperature, and pressure. If the unit needs a defrost cycle to clear frost or ice from the outdoor coil, the defrost initiation control should start a defrost cycle.

If a defrost cycle is called for, and if the defrost relay checks out as working, then the initiation control should be checked. The initiation device is basically an electric switch. When defrost is needed, the switch closes and completes an electric circuit. This circuit is the one that energizes the defrost relay.

To check the switch, make sure that there is electric power in the circuit. Check the voltage across the switch terminals with a voltmeter. Place the probes from the meter leads on each side of the switch terminals, as shown in Figure 22–9. Observe the meter. It should show normal control voltage if the switch is open.

If the voltmeter shows that there is power to the switch, use a heavy wire jumper and place it across the switch terminals. If the heat pump goes into a defrost cycle with the jumper in place, then the initiation device is not working properly. It should be replaced.

Defrost Termination

Defrost termination is usually controlled by the same control that initiates defrost. If the unit does not come out of defrost when the frost or ice is melted from the outside coil, the initiation/termination device should be checked to see whether it is terminating the cycle when it should.

Use a voltmeter, and check for voltage across the defrost relay coil. Place the leads from the voltmeter on each of the coil terminals. Make sure that the unit is not running but that the electric power is on. There should be no electric power in this circuit if the initiation/termination device is working properly. If there is power in the circuit, the device is faulty and should be replaced.

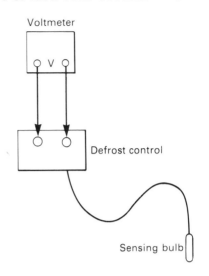

FIGURE 22–9
Checking defrost initiation switch with voltmeter

22.6

Expansion Devices

An air conditioning system must have an **expansion device** to cause a drop in pressure as the refrigerant goes into the evaporator coil. A heat pump, however, acts as both an air conditioner on the cooling cycle and a heating unit on the heating cycle. This dual purpose is accomplished by reversing the flow of the refrigerant in the system. The relative position of the two coils in the system is also reversed when the refrigerant flow is reversed. Thus, an extra expansion device is needed on the condenser coil, which now becomes an evaporator, when the unit is used as a heating unit. Since refrigerant will not flow backward through most expansion devices, a bypass of some kind is also needed around each expansion device.

Expansion device

Bypass

Check valve

Capillary tube expansion devices

As shown in Figure 22–10, many heat pumps use thermal expansion valves with bypasses and check valves as expansion devices. In this system, a valve is located on the liquid line at each coil. A **bypass** is built around each valve in the refrigerant line so that the refrigerant can go around it in one direction. A **check valve** is installed in each bypass to ensure that the refrigerant goes through the valve in the other direction.

The service procedures for checking the other expansion valves for a heat pump are the same as the procedures for an air conditioning system. These procedures were covered in Chapter 17.

Small heat pumps, especially packaged units, often have **capillary tube expansion devices**. A capillary tube is simply a long, small-bore tube that provides enough friction to cause a pressure drop in the liquid refrigerant line. And it will provide that pressure drop regardless of which way the refrigerant flows through it. One capillary tube can be used in the liquid line between the two coils. It will generally have some arrangement for increasing the pressure drop as the refrigerant flows one way and for decreasing the pressure drop as it flows in the other direction. Servicing the capillary tube expansion device in a heat pump is the same as servicing the device in an air conditioning system.

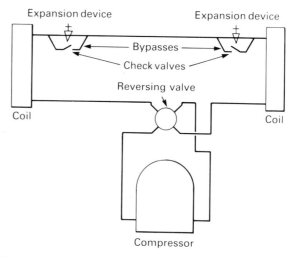

FIGURE 22–10
Expansion valve and bypass arrangement for heat pump

Some units use combination metering valves and check valves as expansion devices. These valves are installed in the liquid line where it connects to each coil. To service these devices, follow the instructions furnished by the manufacturer of the equipment.

Check valves are mechanical devices and are normally not subject to failure. If a check valve fails to open when it should, it causes a high pressure in the refrigerant line ahead of the valve and a low pressure in the line leaving the valve. This pressure differential causes a high pressure drop and consequently a temperature drop in the refrigerant in the line. Checking the temperature of the lines ahead of and behind the check valve should show whether the valve is sticking closed.

22.7

Summary

The service procedures for a heat pump system are, for the most part, the same as the procedures for an air conditioning system. However, because a heat pump unit has reverse-cycle operation when it is in the heating mode, some extra refrigeration and electric control devices are used on it. The most important of these devices are the reversing valve, the thermostat, the defrost controls, and the expansion valve arrangements.

The reversing valve is a mechanical valve that reverses the flow of the refrigerant when the thermostat calls for the heating mode. Most reversing valves use pilot valves. When checking a reversing valve, first determine whether the electric controls that energize the valve are functioning. Then, check the solenoid that controls the pilot valve. Finally, check the mechanical function of the valve.

A heat pump thermostat provides several functions. It must control the reversing valve, control the compressor for both cooling and heating, and control the auxiliary heater during the heating mode. Checking a heat pump thermostat is similar to checking a regular cooling system thermostat. The wiring diagram for the unit that the thermostat is used on must be studied to determine what type of control it is expected to furnish.

Auxiliary heat is used with heat pumps to provide the required amount of heat to a building at colder outside temperatures. The control system for the auxiliary heat is a necessary and integral part of the total control system. A serviceperson should be able to service the auxiliary heat controls as well as the other controls on the heat pump system.

Defrost controls on a heat pump are used to remove any frost or ice that collects on the outdoor coil. The control is provided by one or more defrost relays. These controls are checked by using a voltmeter or an ohmmeter. The defrost initiation and termination cycles must also be checked if they are not functioning properly.

Small heat pumps use several different types of expansion devices. Some of the most common devices are combination metering valves and check valves, combinations of thermal expansion valves and check valves, and capillary tubes with special arrangements for added pressure drop in one mode or the other. All of these devices are checked and serviced in the same way that similar valves are checked in air conditioning systems.

Heat pump servicepersons should be able to check the special components and parts found in heat pumps. They should also be able to check those parts that are common to both air conditioning and heat pump systems.

22.8

Questions

1. Name the parts of the heat pump system for which the reversing valve has refrigerant line connections.
2. What is meant by the following statement? A reversing valve is a pilot-operated valve.
3. What positions the pilot valve piston on a reversing valve?
4. What makes the main piston cylinder move when a reversing valve changes from one mode to the other?

5. Is the following statement true or false? Checking the voltage at the control terminal connections on a solenoid indicates whether the solenoid is good or bad.

6. Is the following statement true or false? In the cooling mode, the temperature of the refrigerant line running from the reversing valve to the indoor coil will be warmer than the temperature of the line running to the outdoor coil.

7. What three things happen when a defrost control calls for defrost?

8. What is indicated if the outdoor coil ices up but the unit does not go into a defrost cycle?

9. Why does a heat pump system need some way to provide refrigerant expansion at the coil at each end of the liquid line?

10. Is the following statement true or false? A check valve allows refrigerant to flow in either direction.

23 Maintenance

23.1

Introduction

Air conditioning and heat pump systems consist of many mechanical components and parts. These mechanical parts need proper care and maintenance to prevent excessive wear. Maintenance of these systems not only ensures that they will last longer but also prevents many emergency breakdowns of the equipment.

Troubleshooters and servicepeople should be able to provide regular maintenance as well as emergency service to air conditioning and heat pump equipment. This chapter describes the details of a maintenance program and the maintenance procedures for specific mechanical parts.

23.2

Regular Maintenance Program

Air conditioning and heat pump systems have many mechanical parts and electric controls and devices. All mechanical parts and most electric components require regular maintenance to prevent excessive wear. Thus, all air conditioning and heat pump systems should have regular maintenance. The **maintenance program** should provide inspection, cleaning, lubrication, necessary adjustments, and any general work required for the continued and proper operation of the system. Maintenance calls should be scheduled on a regular basis. And each

Maintenance program

call should be planned to take care of whatever maintenance is necessary at the time of the call.

Maintenance can be performed by a troubleshooter or a service technician. Or it may be performed by a maintenance technician who is specially trained for performing necessary maintenance procedures but not for performing service work.

23.3

Maintenance Calls

A **maintenance call** should include inspection, a general check of the equipment, and some very specific checks of parts. In the inspection, a visual examination should be made of the equipment, the components, and the parts for dirt or soil that may affect the way the system works. This visual inspection should include looking for dirty coils on the condenser and evaporator and looking for frost or ice on the coils. Also, the electric contacts on contactors, starters, and relays should be inspected for dirt or dust that will affect the way they work. If any dirt, dust, grease, or accumulation of any material that will adversely affect the operation of the equipment is found, it should be cleaned off carefully. Care should be taken to not damage any equipment in the cleaning process.

A general inspection should also include examining the electric wiring in the unit. If any breaks are found in the insulation on the wiring, or if loose terminal connections are found, the wire should be replaced and the connections should be tightened.

In addition, a general inspection should include observation of the system while it is operating. This examination will show whether all of the major components are working.

Finally, a visual inspection of the individual parts should be made. These parts include the filters, the coils, the motors, the blowers, the belts, and any other part that may require special care. If any part is not operating as it should, a more extensive inspection is called for. This more extensive inspection may have to be made by a troubleshooter or a serviceperson who is qualified to determine what is wrong with the part.

Maintenance call

23.4

Maintenance of Specific Parts

Several specific parts of a system must be checked during a maintenance call. These parts include air filters, blower belts, and bearings. The following subsections describe the service and maintenance procedures for these parts.

Air Filters

Air filters

If the **air filters** in a system get plugged with dust or dirt, the amount of air circulating through the system is reduced. If there is not enough air moving across the evaporator coil in an air conditioning system, or across the indoor coil in a heat pump system, abnormally high head pressures result. The high pressures make the unit operate less efficiently and eventually cause the compressor to fail.

The air filters in a system are located in the return air duct or the blower cabinet. They should always be located upstream from the blowers and the coils. An access panel is installed in the ducts or in the equipment cabinet where the filters are located. Figure 23–1 illustrates the typical location of the filters in a system.

To inspect the filters, turn off the electric power to the unit so that the blower does not come on. Inspect the filters visually while they are in place. Or you may

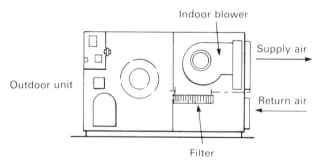

FIGURE 23–1
Location of air filter in typical heat pump system

remove them from the unit to check them. A visual inspection is normally all that is required to see whether the filters are dirty.

If an excessive amount of dust or dirt is found on or in the filter material, the filters should be removed and cleaned. If the filters are covered with dust, the dust can be shaken out by striking them against something solid. If the filter is a cleanable type, such as the filter shown in Figure 23–2, the dust or dirt may be removed by blowing it out with compressed air or by washing it away with water. If the filter is a throwaway type, such as the filter shown in Figure 23–3, it should be replaced with a new, clean filter.

Some air filters have filter material that is more or less permanent. These filters can be cleaned by washing them with water from a hose. Figure 23–4 shows such a permanent filter.

After the filters are cleaned or replaced, the power should be turned on to place the system back in service.

FIGURE 23–2
Cleanable air filter (Courtesy of The American Air Filter Co.)

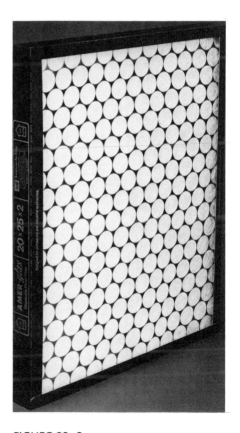

FIGURE 23–3
Throwaway air filter (Courtesy of The American Air Filter Co.)

Blower belts

Bearings

FIGURE 23–4
Permanent air filter (Courtesy of The Farr Co.)

Blower Belts

During a maintenance inspection, all of the **blower belts** in the system should be checked for wear, tension, and alignment. If belts appear to be worn, they should be replaced with new belts to prevent an emergency breakdown. Belts become frayed if they are not aligned correctly or if they are too tight or too loose. They become damaged if they rub against any mechanical part of the system.

To check a belt—and, consequently, the pulleys it runs on—for alignment, first turn off the power so that the motor does not come on. Then, lay a straightedge along the side of the belt and the pulleys. Observe the belt and the pulleys. The pulleys should be lined up with the belt. If they are out of alignment, as shown in Figure 23–5, loosen one or both of the set screws that hold the pulleys in place. Move one or both of the pulleys in or out on their shafts until the pulleys and the belt are in a straight line. Make sure that the pulleys and belt do not rub against any part of the equipment. After the pulleys and belt are aligned, tighten the set screws that hold the pulleys on the shafts. Turn the power to the unit back on.

A belt has the proper tension when it can be deflected approximately 1 inch with a reasonable amount of pressure. To check for tension, turn off the power to the unit. Then, lay a straightedge along the top of the belt from one pulley to the other. With your hand, pull the belt on one side of the pulleys toward the other side. Measure the deflection from the straightedge to the belt. When the resistance to deflection becomes noticeable, there should be about 1 inch of space between the straightedge and the belt, as indicated in Figure 23–6.

If there is more or less than 1 inch of deflection in the belt, adjust the tension on the belt. Loosen the bolts holding the motor in its mount, and move the motor back or forth until the proper tension is achieved. Tighten the motor mount bolts, and turn the power back on.

Bearings

All shaft **bearings** should be checked during a maintenance inspection. Make sure that the bearings are not getting worn. Also, check to see whether they have sufficient lubrication. Blower motors have bearings in the end bells where the shaft runs. Blowers have bearings where the blower wheel shaft is mounted to the blower frame. These bearings should be inspected regularly and should be lubricated when necessary.

A bearing inspection involves three checks. A visual inspection is made to determine whether any sign of damage is present. A check of the bearing temperature is made while the unit is running or immediately after it is turned off. And a physical inspection is made to determine whether there is excessive play in the shaft.

When checking a bearing temperature, carefully touch it with your hand. If it is warmer than the ambient temperature, or if it is hot to the touch, it may need lubrication or even replacement. Attempt to move the shaft in the bearing. There should be no movement at a right angle to the axis of the shaft and very little lateral movement. Excessive movement indicates that the bearing is worn, and it should be replaced.

Bearings need to be lubricated regularly—not too often, but regularly. Older motors and blowers, especially large ones, may have grease cups or grease gun fittings on the bearings. All grease cups should be checked during an inspection to make sure that they have grease in them. If the unit has grease gun fittings, each fitting should be greased on a regular basis. Always follow the manufacturer's recommendations for how often the bearings should be lubricated.

Some bearings have oil cups, which are small cups with hinged lids. Lubrication of these bearings is accomplished by raising the lid and placing oil in the cup. Most motors only need to be lubricated about once a year. And they need only two or three drops of oil then. But again, follow the bearing manufacturer's directions.

Bearings in new units may not have any provisions for lubrication. In the past fifteen to twenty years, many bearings have been manufactured from porous metals that are impregnated with grease in the manufacturing process. Consequently, they never need oiling.

If any other mechanical devices on the system require lubrication, each device should be inspected on each maintenance call. Oil or grease should be applied as needed.

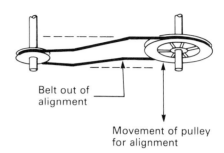

FIGURE 23–5
Alignment of pulleys and belts

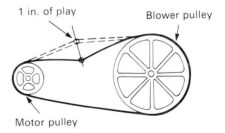

FIGURE 23–6
Adjusting belt tension

23.5

Record Keeping

Maintenance record

A record should be kept for every maintenance job. The person making a maintenance call should make a complete list of what is checked on the call. The **maintenance record** should include details of the call, such as when it was made, what was done, and any notes pertaining to what should be checked on the next call. Some maintenance procedures need to be performed on every visit. Others may only need to be performed on a semiannual or annual basis. An accurate record of what is done on each visit allows the maintenance person to determine what needs to be taken care of on each visit.

Visual inspections and routine checks, such as inspections of the filters and of the condition of the electric wiring, should be performed on every call. Checking the operation of the controls may only be required at the beginning of each operational season. Lubrication of

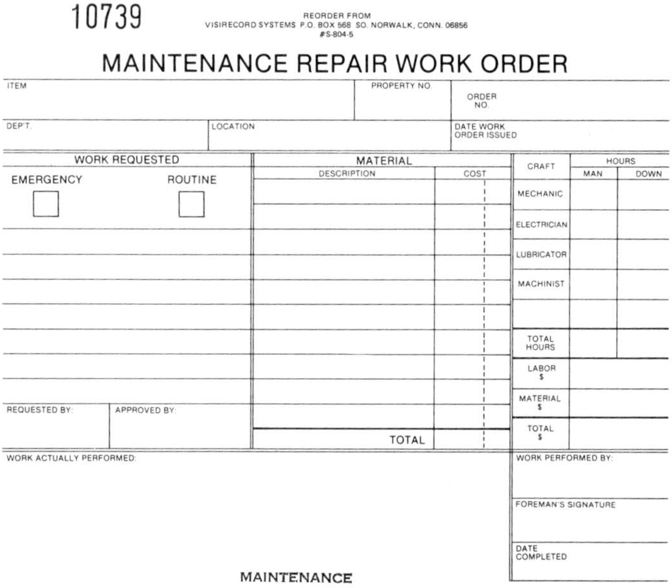
10739

REORDER FROM
VISIRECORD SYSTEMS P.O. BOX 568 SO. NORWALK, CONN. 06856
#S-804-5

MAINTENANCE REPAIR WORK ORDER

ITEM | PROPERTY NO. | ORDER NO.

DEP'T. | LOCATION | DATE WORK ORDER ISSUED

WORK REQUESTED: EMERGENCY ☐ ROUTINE ☐

MATERIAL: DESCRIPTION | COST | TOTAL

CRAFT | HOURS: MAN | DOWN
MECHANIC
ELECTRICIAN
LUBRICATOR
MACHINIST
TOTAL HOURS
LABOR $
MATERIAL $
TOTAL $

REQUESTED BY: | APPROVED BY:

WORK ACTUALLY PERFORMED:

WORK PERFORMED BY:
FOREMAN'S SIGNATURE
DATE COMPLETED

MAINTENANCE

A. Report

FIGURE 23–7 Maintenance report (A) and record (B) (Courtesy of Visirecords) blower motors is normally only necessary on an annual basis.

The person performing the maintenance should fill out a report form on the job as the work is done. A permanent record should also be kept for the job in a master file at the office. Figure 23–7 illustrates a typical report form to be filled out by the maintenance person (Figure 23–7A) and a master record to be kept at the office (Figure 23–7B). Properly filled out maintenance report forms and permanent records of calls made make it easier for a maintenance person to do a good job. They also reduce callbacks and emergency service calls.

23.6

Summary

Air conditioning and heat pump systems contain many mechanical and electric parts. These parts require reg-

PREVENTIVE MAINTENANCE WORK ORDER			OUT	OUT	OUT					
ITEM				PROPERTY NO.	DATE ISSUED					
SERIAL NO.	DEPARTMENT	LOCATION			DATE COMPLETED					

FREQ. CODE	
1	ANNUAL
2	SEMI-ANNUAL
4	QUARTERLY
6	BI-MONTHLY
12	MONTHLY
26	SEMI-MONTHLY

NOTE: DO ENCIRCLED OPERATION NUMBERS ONLY

OPER. NO.	FREQ. CODE	PREVENTIVE MAINTENANCE OPERATION	√ - OK X - SEE NOTE	STANDARD

PREV. MAIN. SCHEDULE				WEEK	
MISC.	LUB.	ELEC.	MECH.	RED	BLUE
				1	27
				2	28
				3	29
				4	30
				5	31
				6	32
				7	33
				8	34
				9	35
				10	36
				11	37

B. Office Record

ular maintenance. Proper maintenance of heat pump and air conditioning systems ensures that the equipment performs as it should. Maintenance also prevents some emergency service because of equipment failure. A maintenance call, unlike a troubleshooting or a service call, is a regular call made to ensure the proper operation of the equipment. It is not initiated by a complaint concerning the operation of the system.

A maintenance call includes both visual and mechanical checks of the general system, the components, and the individual parts of the system. The general operation of the equipment is checked during a normal operational cycle. If the system is an air conditioning system, this check involves the cooling cycle only. If the system is a heat pump system, then both the heating and the cooling modes should be checked.

Power wiring, control wiring, controls, and operating parts are checked regularly to make sure that each is operating as it should. Filters are checked, cleaned, and changed, if necessary. Blower belts and motors are checked for proper operation. Blower wheel bearings and motor bearings are checked and lubricated when necessary.

A complete record of each maintenance call is made at the time of the call. A master file with all pertinent data concerning the system is maintained in the office to keep track of what needs to be done on each call.

Proper maintenance of air conditioning and heat pump equipment will prolong the operating life of the equipment. It will also reduce the number of emergency service calls required for the system. As a result, the owner will have less trouble with the equipment, and the total cost of the system will be reduced.

23.7

Questions

1. Is the following statement true or false? After an air conditioning or a heat pump system is installed, it never needs any care.
2. A maintenance program ensures that equipment will _____ longer and also prevents _____ breakdowns.

3. Name four important elements of a maintenance program.
4. How do dirty air filters affect the operation of a system?
5. Is the following statement true or false? All filters can be washed out with water and reinstalled.
6. How can you check the pulleys on a belt-driven blower for proper alignment?
7. Is the following statement true or false? All bearings require lubrication during maintenance calls.
8. How can you tell whether a bearing is lubricated for its lifetime?
9. Why should a record be kept of what is performed on each maintenance call and on each service call to a system?

Appendix Tables

TABLE A
Air Conditioning Troubleshooting Chart

Complaint	Symptom	Cause	Section, component, or part
No cooling	No air coming from supply register	No electric power to equipment	Electric power supply system
		Control system not functioning correctly	Operating controls Safety controls
		Evaporator blower not operating	Evaporator section
		Evaporator coil obstructed	
		Dampers closed	Air distribution system
	Warm air coming from supply registers	System undersized	Entire system
		Controls not functioning correctly	Controls
		Evaporator not operating	Evaporator blower Expansion device Evaporator coil
		Compressor not operating	Condensing section
		Condenser blower not operating	
		Condensing coil obstructed	
		Return air too cool	Return grille location
		Ductwork not insulated	Ductwork
Not enough cooling	Air from supply registers too warm	Compressor cycling	Condensing unit
		Condenser blower not working	
		Noncondensable gases in condenser coil	
		Dirty condenser coil	
		Evaporator blower turning too slowly	Evaporator section
		Obstructed or dirty evaporator coil	
		Expansion device faulty	Expansion device
		Supply ductwork not insulated in warm area	Air distribution system

TABLE A
Air Conditioning Troubleshooting Chart (*Continued*)

Complaint	Symptom	Cause	Section, component, or part
	Room air temperature warmer than set point of thermostat	Equipment too small Thermostat operation Thermostat calibration	Equipment in general
		Anticipator setting	Controls
		Dirty or obstructed evaporator coil	Evaporator section
		Dirty filters Ductwork undersized	Air distribution system
Too cool	Room temperature cooler than thermostat set point	Equipment too large	Entire system
		Thermostat operation Thermostat calibration Control relay	Operating controls
	Cool drafts in room or area	Equipment oversized	Equipment in general
		Supply registers poorly located	Air distribution system
		Wrong type of registers for job	
		System never balanced	
Noisy system	Popping or ticking noises from registers	Metal ductwork not braced properly	Air distribution system
	Scraping or squeaking noises from registers	Evaporator blower wheel rubbing on housing	Evaporator section
		Blower bearings worn out	
		Blower motor bearings worn out	
		Blower belt slipping	
	Noisy compressor	Refrigerant flooding back to compressor	Expansion device Evaporator blower Refrigerant charge
		Compressor bearings worn out	Compressor
		Broken valves in compressor	
		Low oil level in compressor	

TABLE A
Air Conditioning Troubleshooting Chart (*Concluded*)

Complaint	Symptom	Cause	Section, component, or part
	Noisy condensing unit	Condenser blower blades hitting housing	Condensing section
		Condenser blower bearings worn out	
		Condenser blower motor bearings worn out	
	Noisy evaporator blower	Blower blades striking against blower housing	Evaporator blower
	Noise from refrigerant lines where they go through building	Lines not insulated from structure	Refrigerant lines
		No vibration isolator on hot gas discharge line	
Too expensive to operate	Unit runs constantly	Equipment undersized for job	Total system
		Thermostat out of calibration	Control system
		Thermostat contracts welded shut	
		Control circuit wiring short-circuited	
		Control relay contacts welded shut	
		Evaporator blower not operating	Evaporator section
		Low refrigerant pressure on low side	
		Dirty evaporator coil	
		Condenser blower not operating	Condenser section
		Dirty filters	Air distribution system
	Unit needs frequent service	System improperly designed	Total system
		System improperly installed	
		Refrigerant system not properly cleaned up after compressor burnout	
		Inadequate maintenance	

TABLE B
Heat Pump Troubleshooting Chart for Cooling or Heating

Complaint	Symptom	Cause	Section, component, or part
Poor temperature control	No air coming from supply registers	No electric power to equipment	Electric power supply system
		Controls not functioning properly	Operating controls Safety controls
		Indoor blower not operating	Indoor blower
		Indoor coil obstructed or dirty	Indoor coil
		Dampers closed	Air distribution system
	Not enough air from supply registers	Indoor blower motor cycling on overload Indoor blower belts slipping	Indoor blower
		Dirty filters Obstructed or dirty indoor coil	Indoor unit
		Ductwork, registers, or grilles too small Dampers partially closed	Air distribution system
	Room temperature does not match set point of thermostat	Equipment oversized or undersized	Equipment in general
		Thermostat out of calibration Anticipator set incorrectly on thermostat	Operating controls
Noisy system	Popping or ticking noises from register	Metal ductwork not braced properly	Air distribution system
	Scraping or squeaking noises from registers	Indoor blower wheel rubbing on housing Blower bearings worn out Blower motor bearings worn out Blower belt slipping	Indoor blower
	Noisy compressor	Refrigerant flooding back to compressor	Expansion device Refrigerant charge
		Compressor bearings worn out	Compressor

TABLE B
Heat Pump Troubleshooting Chart for Cooling or Heating (*Concluded*)

Complaint	Symptom	Cause	Section, component, or part
	Noisy outdoor unit	Broken valves in compressor	
		Low oil level in compressor	
		Outdoor blower blades hitting housing	Outdoor section
		Outdoor blower bearings worn out	
		Outdoor blower motor bearings worn out	
	Noisy indoor unit	Indoor blower blades hitting housing	Indoor section
	Noise from refrigerant lines where they go through building	Lines not insulated from structure	Refrigerant lines
		No vibration isolator on hot gas discharge line	
Too expensive to operate	Unit runs constantly	Equipment undersized	Entire system
		Thermostat out of calibration	Control system
		Thermostat contacts welded shut	
		Control circuit wiring short-circuited	
		Control relay contacts welded shut	
		Outdoor unit blower not operating	Outdoor unit
		Dirty indoor coil	Indoor unit
		Auxiliary heat controls stuck	Auxiliary heat control system
		Unit will not come out of defrost	Defrost control system
		Dirty filters	Air distribution system
	Unit needs frequent service	System improperly designed	Total system
		System improperly installed	
		System not cleaned up properly after a burnout	
		Inadequate maintenance	

TABLE C
Heat Pump Troubleshooting Chart for Cooling Only

Complaint	Symptom	Cause	Section, component, or part
No cooling	No air coming from supply registers	No electric power to equipment	Electric power supply system
		Control system not functioning properly	Operating controls Safety controls
		Indoor blower not operating	Indoor unit
		Indoor coil obstructed or dirty	
		Duct dampers closed	Air distribution system
	Warm air coming from supply registers	System undersized	Entire system
		Controls not functioning properly	Control system
		Indoor coil dirty or obstructed	Indoor unit
		Compressor not operating	Compressor
		Compressor cycling on overloads	
		Outdoor blower not operating	Outdoor unit
		Outdoor coil obstructed	
		Return air too cool	Return air grille location
		Duct not insulated	Duct system
Not enough cooling	Warm air coming from supply registers	System undersized	Entire system
		Operating or safety controls	Controls
		Compressor cycling on overload	Compressor controls
		Outdoor blower not operating	Outdoor unit
		Outdoor coil dirty or obstructed	
		Dirty filters	Indoor unit
		Indoor coil dirty	
		Indoor blower belt slipping	

TABLE C
Heat Pump Troubleshooting Chart for Cooling Only (*Concluded*)

Complaint	Symptom	Cause	Section, component, or part
		Return air grilles in cold area	Air distribution system
		Supply ductwork not insulated	
	Room air temperature warmer than set point of thermostat	Equipment undersized	Entire system
		Thermostat operation	Controls
		Thermostat calibration	
		Anticipator setting	
		Indoor coil dirty or obstructed	Indoor unit
		Dirty filters	Air distribution system
		Ductwork undersized	
		Supply registers too small	
		Return air grilles too small	
Too cool	Room temperature lower than thermostat set point	Equipment in general too large	Entire system
		Thermostat operation	Controls
		Thermostat calibration	
		Anticipator setting	
	Cool drafts in room or area	Equipment oversized	Equipment in general
		Supply registers poorly located	Air distribution system
		Wrong type of registers for job	
		System never balanced	

TABLE D
Heat Pump Troubleshooting Chart for Heating Only

Complaint	Symptom	Cause	Section, component, or part
No heat	No air coming from supply registers	No electric power to equipment	Electric power supply system
		Controls not functioning properly	Operating controls Safety controls
		Indoor blower not operating Indoor coil obstructed	Indoor unit
		Duct dampers closed	Air distribution system
	Cool air coming from supply registers	Unit stays in cooling cycle	Thermostat Control system Reversing valve
		Unit will not come out of defrost	Defrost controls
		Auxiliary heat does not come on	Auxiliary heat controls
	Room air temperature lower than set point of thermostat	Thermostat operation Thermostat calibration Anticipator setting	Controls
		Auxiliary heat unit undersized	Auxiliary heat unit
		Auxiliary heat not coming on	Auxiliary heat controls
		Dirty filters Obstructed indoor coil	Indoor unit
		Ductwork undersized Return air grilles too small	Air distribution system
Not enough heat	Supply air too cool	Compressor cycling on overloads	Controls
		Indoor blower runs continuously Dirty filters Dirty indoor coil	Indoor unit
		Outdoor blower not working Dirty outdoor coil	Outdoor unit
		Supply ductwork not insulated in cold area	Air distribution system

TABLE D
Heat Pump Troubleshooting Chart for Heating Only (*Concluded*)

Complaint	Symptom	Cause	Section, component, or part
	Room temperature cooler than thermostat set point	Thermostat operation Thermostat calibration Anticipator setting	Controls
Too hot	Temperature in building warmer than set point of thermostat	Thermostat operation Thermostat calibration Anticipator setting	Controls

TABLE E
Air Conditioning Service Chart

Complaint	Problem	Major section	Component	Part	Reference
No cooling	No air coming from supply registers	Electric power to unit	Main power supply	Branch: disconnect	Chap. 9
				Wiring	Chap. 9
				Fusing or breakers	Chap. 9
				Unit: disconnect	Chap. 9
				Wiring	Chap. 9
				Fusing or breakers	Chap. 9
			Control circuit	Transformer	Chap. 10
				Control wiring	Chap. 10
				Thermostat	Chap. 10
				Control relay	Chap. 9
				Compressor motor start relay	Chap. 12
				Motor capacitors	Chap. 12
				Compressor overloads	Chap. 10
				High- or low-pressure switch	Chap. 10
		Low-pressure side of system	Evaporator section	Restricted coil	Chap. 17
				Iced-up coil	Chap. 17
			Expansion device	Blocked screen	Chap. 17
				Remote bulb	Chap. 17
				Power head	Chap. 17
				Plugged cap tube	Chap. 17
		Air distribution system	Evaporator blower	Motor	Chap. 12
				Motor relay	Chap. 10
				Motor overloads	Chap. 10
			Dampers	Register dampers	Chap. 18
				Balancing dampers	Chap. 18
				Fire dampers	Chap. 18

TABLE E
Air Conditioning Service Chart (*Continued*)

Complaint	Problem	Major section	Component	Part	Reference
	Warm air coming from supply registers	General	Equipment	System undersized	
			Refrigerant	Overcharge	Chap. 17
				Low charge	Chap. 17
		Controls	Control circuit	Transformer	Chap. 10
				Wiring	Chap. 10
				Thermostat	Chap. 10
				Control relay	Chap. 9
		High-pressure side of system	Compressor	Motor	Chap. 12
				Valves	Chap. 15
				Start switch	Chap. 10
				Capacitor	Chap. 12
			Condenser section	Blower motor	Chap. 12
				Blower relay	Chap. 10
				High-pressure switch	Chap. 10
				Restricted coil	Chap. 15
				Noncondensable gases in coil	Chap. 15
		Low-pressure side of system	Evaporator	Low-pressure switch	Chap. 10
				Obstructed coil	Chap. 17
				Iced-up coil	Chap. 17
			Expansion device	Remote bulb	Chap. 17
				Power head	Chap. 17
				Superheat adjustment	Chap. 17
		Distribution system	Blower	Blower motor	Chap. 12
				Slipping belt(s)	Chap. 18
				Wrong rotation	Chap. 18
				Motor capacitor	Chap. 12
			Filters	Dirty	Chap. 18
				Restricted	
			Ductwork	Uninsulated ducts in warm area of building	Chap. 18

TABLE E
Air Conditioning Service Chart (*Continued*)

Complaint	Problem	Major section	Component	Part	Reference
Not enough cooling	Warm air coming from supply registers	General	Equipment	System undersized	
			Refrigerant	Overcharge	Chap. 17
				Undercharge	Chap. 17
		Controls	Control circuit	Thermostat	Chap. 10
				Anticipator	Chap. 10
		High-pressure side of system	Compressor	Motor overloads	Chap. 12
			Condensing unit	Blower motor	Chap. 12
				Blower relay	Chap. 10
				High- or low-pressure switches	Chap. 10
				Restricted coil	Chap. 15
				Noncondensable gases in coil	Chap. 15
		Low-pressure side of system	Evaporator	Blower motor	Chap. 12
				Blower relay	Chap. 10
				Restricted or dirty coil	Chap. 18
				Iced-up coil	Chap. 17
			Expansion device	Remote bulb	Chap. 17
				Power head	Chap. 17
				Superheat adjustment	Chap. 17
		Distribution system	Blower	Blower motor	Chap. 12
				Slipping belt(s)	Chap. 18
				Wrong rotation	Chap. 18
				Motor capacitor	Chap. 12
			Filters	Dirty	Chap. 18
				Restricted	Chap. 18
			Ductwork	Uninsulated ducts in warm part of building	Chap. 18
	Cannot get room temperature as low as thermostat set point	General	Equipment in general	Undersized	

TABLE E
Air Conditioning Service Chart (*Continued*)

Complaint	Problem	Major section	Component	Part	Reference
			Refrigerant	Overcharge	Chap. 17
				Undercharge	Chap. 17
		Controls	Control circuit	Thermostat	Chap. 10
				Anticipator	Chap. 10
		High-pressure side of system	Compressor	Motor	Chap. 12
				Motor overloads	Chap. 10
				High-pressure control	Chap. 10
				Motor starting relay	Chap. 12
				Capacitors	Chap. 12
				Valves	Chap. 15
			Condenser section	Blower motor	Chap. 12
				Blower drives	Chap. 18
				Dirty coil	Chap. 15
		Low-pressure side of system	Evaporator section	Restricted coil	Chap. 17
				Restricted distributor	Chap. 17
			Expansion device	Restricted	Chap. 17
				Power head	Chap. 17
				Remote bulb	Chap. 17
				Superheat setting	Chap. 17
			Filter drier	Restricted	Chap. 17
			Refrigerant lines	Pinched or restricted	Chap. 15
		Air distribution system	Blower	Blower sizing	Chap. 18
				Blower motor	Chap. 12
				Blower overloads	Chap. 10
				Blower speed	Chap. 18
				Direction of rotation	Chap. 18
			Filters	Dirty	Chap. 18
				Restricted	Chap. 18
			Ductwork	Undersized	Chap. 18
				Location	Chap. 18
				Not insulated	Chap. 18
				Air leaks	Chap. 18
			Registers and grilles	Sizing	Chap. 18
				Location	Chap. 18

TABLE E
Air Conditioning Service Chart (*Continued*)

Complaint	Problem	Major section	Component	Part	Reference
Too cool	Room temperature stays lower than thermostat setting	General	Unit too large for job	Entire system	
		Controls	Thermostat	Location	Chap. 10
				Level	Chap. 10
				Calibration	Chap. 10
	Cool drafts in rooms	Entire system	Equipment	All parts	
		Air distribution system	Supply registers	Location	Chap. 18
				Type	Chap. 18
				Sizing	Chap. 18
Noisy system	Popping or ticking noises from supply registers	Air distribution system	Supply ducts	Hangers	Chap. 18
				Stiffeners	Chap. 18
	Scraping or squeaking noises	Low-pressure side of system	Evaporator blower	Blower wheel	Chap. 18
				Blower pulleys	Chap. 18
				Bearings	Chap. 18
				Motor	Chap. 12
	Noisy compressor	High-pressure side of system	Compressor	Bearings	Chap. 15
				Valves	Chap. 15
				Pistons or crank assembly	Chap. 15
				Oil level	Chap. 15
				Incorrect refrigerant piping	Chap. 15
	Noisy condensing unit	High-pressure side of system	Condenser blower	Wheel or fan	Chap. 15
				Bearings	Chap. 15
				Motor bearings	Chap. 12
	Noisy evaporator unit	Low-pressure side of system	Evaporator blower	Blower wheel	Chap. 18
	Refrigerant lines	High-pressure side of system	Hot gas discharge and liquid lines	Line hangers	Chap. 18
				Vibration isolators	Chap. 18
				Insulation	Chap. 18
Too expensive to operate	Unit runs constantly	General	Unit	Equipment undersized for job	

TABLE E
Air Conditioning Service Chart (*Concluded*)

Complaint	Problem	Major section	Component	Part	Reference
		Controls	Operating controls	Thermostat	Chap. 10
				Control relay	Chap. 9
		High-pressure side of system	Condenser	Condenser coil dirty	Chap. 15
		Low-pressure side of system	Evaporator	Evaporator blower motor	Chap. 12
				Evaporator coil dirty	Chap. 17
				Low refrigerant charge	Chap. 17
		Air distribution system	Ductwork	Dirty filters	Chap. 18
				Ducts undersized	Chap. 18

TABLE F
Heat Pump Service Chart for Cooling or Heating

Complaint	Problem	Major section	Component	Part	Reference
No cooling or heating	No air coming from supply registers	Electric power	Power supply	Branch: disconnect	Chap. 9
				Wiring	Chap. 9
				Fusing or breakers	Chap. 9
				Unit: disconnect	Chap. 9
				Wiring	Chap. 9
				Fusing or breakers	Chap. 9
			Control circuit	Transformer	Chap. 10
				Wiring	Chap. 10
				Thermostat	Chap. 22
				Control relay	Chap. 9
				Motor start relay	Chap. 12
				Motor overloads	Chap. 10
				Capacitors	Chap. 12
				High- or low-pressure switch	Chap. 10
		Indoor section of system	Indoor coil	Restricted	Chap. 17
		Air distribution system	Indoor unit blower	Blower motor	Chap. 12
				Blower drives	Chap. 18
				Motor overloads	Chap. 10
			Dampers	Register dampers	Chap. 18
				Balancing dampers	Chap. 18
				Fire dampers	Chap. 18
	Not enough air from supply registers	Indoor unit	Indoor blower	Blower motor	Chap. 12
				Blower motor controls	Chap. 12
				Blower belt(s)	Chap. 18
			Indoor coil	Dirty coil	Chap. 17
		Air distribution system	Ductwork	Filter	Chap. 18
				Ducts, registers, or grilles too small	Chap. 18

TABLE F
Heat Pump Service Chart for Cooling or Heating (*Continued*)

Complaint	Problem	Major section	Component	Part	Reference
				Dampers partially closed	Chap. 18
	Room temperature does not match set point of thermostat	General	Entire system	Equipment undersized or oversized	
		Controls	Control circuit	Thermostat	Chap. 22
				Anticipator	Chap. 10
				Control relay	Chap. 9
Noisy system	Popping or ticking noises from supply registers	Air distribution system	Supply ducts	Hangers	Chap. 18
				Stiffeners	Chap. 18
	Scraping or squeaking noises	Indoor section of system	Indoor unit blower	Blower wheel	Chap. 18
				Blower pulleys	Chap. 18
				Bearings	Chap. 18
				Motor	Chap. 12
	Noisy outdoor unit	Outdoor section of system	Compressor	Bearings	Chap. 15
				Valves	Chap. 15
				Pistons or crank assembly	Chap. 15
				Oil level	Chap. 15
				Incorrect refrigerant piping	Chap. 15
			Outdoor blower	Wheel or fan	Chap. 15
				Bearings	Chap. 15
				Motor bearings	Chap. 12
	Noisy indoor unit	Indoor section of system	Blower	Motor	Chap. 12
				Wheel	Chap. 18
				Bearings	Chap. 18
	Refrigerant lines	Outdoor section of system	Refrigerant lines	Line hangers	Chap. 18
				Vibration isolators	Chap. 18
				Insulation	Chap. 18
Too expensive to operate	Unit runs constantly	General	Unit	Equipment undersized for job	
		Controls	Operating controls	Thermostat	Chap. 22
				Control relay	Chap. 9

TABLE F
Heat Pump Service Chart for Cooling or Heating (*Concluded*)

Complaint	Problem	Major section	Component	Part	Reference
			Defrost controls	Defrost initiation	Chap. 22
				Defrost termination	Chap. 22
				Defrost relay	Chap. 22
		Outdoor section of system	Outdoor coil	Outdoor coil dirty	Chap. 15
		Indoor section of system	Indoor coil	Indoor coil dirty	Chap. 17
				Low refrigerant charge	Chap. 17
		Auxiliary heat	Controls	Relay	Chap. 22
				Defrost control	Chap. 22
		Air distribution system	Ductwork	Dirty filters	Chap. 18
				Ducts undersized	Chap. 18

TABLE G
Heat Pump Service Chart for Cooling Only

Complaint	Problem	Major section	Component	Part	Reference
No cooling	No air coming from supply registers	Electric power	Electric power supply	Branch: disconnect	Chap. 9
				Wiring	Chap. 9
				Fusing or breakers	Chap. 9
				Unit: disconnect	Chap. 9
				Wiring	Chap. 9
				Fusing or breakers	Chap. 9
			Control circuit	Transformer	Chap. 10
				Wiring	Chap. 10
				Thermostat	Chap. 22
				Control relay	Chap. 9
				Motor start relay	Chap. 12
				Motor overloads	Chap. 10
				Capacitors	Chap. 12
		Indoor section	Indoor blower	Motor	Chap. 12
				Motor relay	Chap. 12
				Belt(s) or drives	Chap. 18
			Indoor coil	Dirty or obstructed coil	Chap. 17
	Warm air coming from supply registers	General	Refrigerant	Overcharge	Chap. 21
				Undercharge	Chap. 21
		Controls	Control circuit	Transformer	Chap. 10
				Wiring	Chap. 10
				Thermostat	Chap. 22
				Control relay	Chap. 9
		Outdoor section of system	Compressor	Motor	Chap. 12
				Valves	Chap. 15
				Start switch	Chap. 10
				Capacitor	Chap. 12
			Outdoor unit	Blower motor	Chap. 12
				Blower relay	Chap. 10
				High-pressure switch	Chap. 10
		Indoor section of system	Indoor unit	Dirty coil	Chap. 17
				Low-pressure switch	Chap. 10

TABLE G
Heat Pump Service Chart for Cooling Only (*Continued*)

Complaint	Problem	Major section	Component	Part	Reference
			Indoor blower	Motor	Chap. 12
				Motor starting relay	Chap. 10
				Capacitors	Chap. 12
				Slipping belt(s)	Chap. 18
				Wrong rotation	Chap. 18
		Distribution system	Filters	Dirty	Chap. 18
				Restricted	Chap. 18
Not enough cooling	Warm air coming from supply registers	General	Equipment in general	System undersized	
		Controls	Operating and safety controls	Thermostat	Chap. 22
				High- or low-pressure switches	Chap. 10
		Outdoor unit	Compressor	Controls	Chap. 10
				Start relay	Chap. 10
				Capacitor(s)	Chap. 12
				Physical damage	Chap. 15
			Condenser blower	Controls	Chap. 10
				Motor	Chap. 12
				Blower wheel or blades	Chap. 15
			Condenser coil	Obstructed coil	Chap. 15
				Nonconden-sable gases	Chap. 15
		Indoor unit	Indoor blower	Controls	Chap. 10
				Motor	Chap. 12
				Start relay	Chap. 10
				Capacitor(s)	Chap. 12
			Indoor coil	Obstructed	Chap. 17
			Expansion device	Blocked	Chap. 17
				Power head	Chap. 17
				Remote bulb	Chap. 17
		Air distribution system	Ducts	No insulation	Chap. 18
			Return air	Return air grilles in a cool location	Chap. 18
	Cannot get room	General	Equipment in general	Undersized	

TABLE G
Heat Pump Service Chart for Cooling Only (*Continued*)

Complaint	Problem	Major section	Component	Part	Reference
	temperature as low as thermostat set point	Controls	Control circuit	Thermostat	Chap. 22
				Anticipator	Chap. 10
		Outdoor section of system	Compressor	Motor	Chap. 12
				Motor overloads	Chap. 10
				High-pressure control	Chap. 10
				Starting relay	Chap. 12
				Capacitors	Chap. 12
				Valves	Chap. 15
			Outdoor coil	Dirty	Chap. 17
				Restricted	Chap. 17
			Outdoor blower	Motor	Chap. 12
				Drives	Chap. 15
				Blower blades	Chap. 15
				Blower relay	Chap. 10
				Capacitor	Chap. 12
				Wrong rotation	Chap. 18
		Indoor section of system	Indoor coil	Dirty coil	Chap. 17
				Low-pressure control	Chap. 10
			Expansion device	Restricted	Chap. 17
				Power head	Chap. 17
				Remote bulb	Chap. 17
				Superheat setting	Chap. 17
		Air distribution system	Indoor blower	Blower sizing	Chap. 18
				Blower motor	Chap. 12
				Blower overloads	Chap. 10
				Blower speed	Chap. 18
				Direction of rotation	Chap. 18
			Filters	Dirty	Chap. 18
				Restricted	Chap. 18
			Ductwork	Sizing	Chap. 18
				Location	Chap. 18
				Not insulated	Chap. 18
				Air leaks	Chap. 18

TABLE G
Heat Pump Service Chart for Cooling Only (*Concluded*)

Complaint	Problem	Major section	Component	Part	Reference
Too cool	Room temperature stays lower than thermostat setting	General	Registers and grilles	Sizing Location	Chap. 18 Chap. 18
			Equipment in general	Too large for job	
	Cool drafts in room	Air distribution system	Supply registers	Location Type Size	Chap. 18 Chap. 18 Chap. 18

TABLE H
Heat Pump Service Chart for Heating Only

Complaint	Problem	Major section	Component	Part	Reference
No heat	No air coming from supply registers	Electric power	Power supply	Branch: disconnect	Chap. 9
				Wiring	Chap. 9
				Fusing or breakers	Chap. 9
				Unit: disconnect	Chap. 9
				Wiring	Chap. 9
				Fusing or disconnect	Chap. 9
			Control circuit	Transformer	Chap. 10
				Wiring	Chap. 10
				Thermostat	Chap. 22
				Control relay	Chap. 9
				Motor start relay	Chap. 12
				Motor overloads	Chap. 10
				Capacitors	Chap. 12
		Indoor section of system	Indoor blower	Motor	Chap. 12
				Control relay	Chap. 9
			Indoor coil	Obstructed or dirty	Chap. 17
	Cool air coming from supply registers	Controls	Control circuit	Transformer	Chap. 10
				Wiring	Chap. 10
				Thermostat	Chap. 22
				Control relay	Chap. 9
				Pressure switches	Chap. 10
			Defrost controls	Termination sensor	Chap. 22
				Termination relay	Chap. 22
		Outdoor section of system	Compressor	Motor	Chap. 12
				Valves	Chap. 15
				Start switch	Chap. 10
				Capacitor	Chap. 12
			Outdoor blower	Blower motor	Chap. 12
				Blower relay	Chap. 9
				Start relay	Chap. 10
				Capacitor(s)	Chap. 12

TABLE H
Heat Pump Service Chart for Heating Only (*Continued*)

Complaint	Problem	Major section	Component	Part	Reference
		Indoor section of system	Indoor coil	Dirty coil	Chap. 17
				Restricted	Chap. 17
			Indoor blower	Motor	Chap. 12
				Start relay	Chap. 10
				Drives	Chap. 18
				Blower relay	Chap. 19
				Wheel or blades	Chap. 18
				Direction of rotation	Chap. 18
		Distribution system	Filter	Dirty	Chap. 18
				Restricted	Chap. 18
			Ductwork	Undersized	Chap. 18
				Location	Chap. 18
				Not insulated	Chap. 18
				Air leaks	Chap. 18
			Registers and grilles	Sizing	Chap. 18
				Location	Chap. 18
	Room temperature stays lower than thermostat setting	General	Equipment in general	Undersized for job	
		Controls	Control circuit	Thermostat	Chap. 22
				Anticipator	Chap. 10
				Calibration	Chap. 10
		Reversing valve	Main valve	Valve piston	Chap. 22
			Pilot valve	Solenoid	Chap. 22
				Valve piston	Chap. 22
		Auxiliary heat	Controls	Thermostat	Chap. 22
				Relay	Chap. 22
				Limits	Chap. 22
			Elements	Wiring	Chap. 9
				Resistance wire	Chap. 22
Not enough heat	Cool drafts in rooms	Air distribution system	Supply registers	Location	Chap. 18
				Type	Chap. 18
				Size	Chap. 18
	Room temperature stays lower than thermostat setting	General	Equipment in general	Undersized for job	
		Controls	Control circuit	Thermostat	Chap. 22
				Anticipator	Chap. 10
				Calibration	Chap. 10

TABLE H
Heat Pump Service Chart for Heating Only (*Concluded*)

Complaint	Problem	Major section	Component	Part	Reference
		Reversing valve	Main valve	Valve piston	Chap. 22
			Pilot valve	Solenoid	Chap. 22
				Valve piston	Chap. 22
		Auxiliary heat	Controls	Thermostat	Chap. 22
				Relay	Chap. 22
				Limits	Chap. 22
			Elements	Wiring	Chap. 9
				Resistance wire	Chap. 22
Too hot	Room temperature stays above thermostat setting	General	Equipment in general	Oversized for job	
		Controls	Control circuit	Thermostat	Chap. 22
				Anticipator	Chap. 10
				Calibration	Chap. 10
		Auxiliary heat	Controls	Thermostat	Chap. 22
				Relay	Chap. 22
				Limits	Chap. 22

Index

Sight glass, 180, 214, 308
Single-phase motor, 165
Single-phase power, 103
Slim-nose pliers, 14
Socket fuse, 108
Socket wrench, 11
Solenoid valve, 131
Solid-state relay, 157, 172
Split phase motor, 151
Starting relay, 155
Starting switch, 155
Sub cooling, 210
Suction accumulator, 246, 267
Suction line, 49, 65, 266
Superheat, 57, 259
Supply register, 73, 282
Symptom, 43, 79
Systematic approach
  service, 91
  troubleshooting, 77
Systematic procedure, 5

Temperature control devices, 242
Temperature drop, 275
Temperature scales
  Celsius, 16
  Fahrenheit, 16
Terminal strip
  high voltage, 129
  low voltage, 129
Test instruments, 15
Thermometers, 15
  dial, 16
  stem, 16
Thermostat, 126, 296, 333
  air conditioning, 126
  heat pump, 296

Thermostatic expansion valve, 242, 258, 261
Three-phase motor, 153, 167
Three-phase power, 103
Time delay fuse, 109
Transformer, 124
  primary winding, 124
  secondary winding, 124
  servicing, 125
  step-down, 124
  step-up, 124
Troubleshooter, 1, 4, 42, 77
Troubleshooting chart, 6, 76, 85

Unit disconnect, 113

Valves, compressor, 201
Valve wrench, 212
Velometer, 283
Vise grip pliers, 14
Voltage, 146
Volume control device, 239

Water pump pliers, 12
Wiring
  continuity, 109
  line side, 102
  load side, 102
Wrench, 9
  adjustable, 10
  box-end, 11
  flare-nut, 11
  open-end, 11
  socket, 11